# Deep Reinforcement Learning for Reconfigurable Intelligent Surfaces and UAV Empowered Smart 6G Communications

## Other related titles:

You may also like

- PBTE111 | Smyth | 6G: Evolution or Revolution? A converged view of cellular, Wi-Fi, computing and communications | in production
- PBPC067 | Duong | Online Optimisation for Real-Time Decision Making: Theory, practice, and real-world applications | contracted May 2022
- PBTE109 | Ahmadi | Digital Twins for 6G: Fundamental theory, technology and applications | pub June 2024
- PBTE107 | Duong | Physical Layer Security for 6G Networks | pub April 2024
- PBTE087 | Nguyen | Real Time Convex Optimisation for 5G Networks | pub Dec 2021

We also publish a wide range of books on the following topics:
Computing and Networks
Control, Robotics and Sensors
Electrical Regulations
Electromagnetics and Radar
Energy Engineering
Healthcare Technologies
History and Management of Technology
IET Codes and Guidance
Materials, Circuits and Devices
Model Forms
Nanomaterials and Nanotechnologies
Optics, Photonics and Lasers
Production, Design and Manufacturing
Security
Telecommunications
Transportation

All books are available in print via https://shop.theiet.org or as eBooks via our Digital Library https://digital-library.theiet.org.

IET TELECOMMUNICATIONS SERIES 106

# Deep Reinforcement Learning for Reconfigurable Intelligent Surfaces and UAV Empowered Smart 6G Communications

Antonino Masaracchia, Khoi Khac Nguyen, Trung Q. Duong and Vishal Sharma

The Institution of Engineering and Technology

## About the IET

This book is published by the Institution of Engineering and Technology (The IET).

We inspire, inform, and influence the global engineering community to engineer a better world. As a diverse home across engineering and technology, we share knowledge that helps make better sense of the world, accelerate innovation, and solve the global challenges that matter.

The IET is a not-for-profit organisation. The surplus we make from our books is used to support activities and products for the engineering community and promote the positive role of science, engineering, and technology in the world. This includes education resources and outreach, scholarships and awards, events and courses, publications, professional development and mentoring, and advocacy to governments.

To discover more about the IET please visit https://www.theiet.org/.

## About IET books

The IET publishes books across many engineering and technology disciplines. Our authors and editors offer fresh perspectives from universities and industry. Within our subject areas, we have several book series steered by editorial boards made up of leading subject experts.

We peer review each book at the proposal stage to ensure the quality and relevance of our publications.

## Get involved

If you are interested in becoming an author, editor, series advisor, or peer reviewer, please visit https://www.theiet.org/publishing/publishing-with-iet-books/ or contact author_support@theiet.org.

## Discovering our electronic content

All of our books are available online via the IET's Digital Library. Our Digital Library is the home of technical documents, eBooks, conference publications, real-life case studies, and journal articles. To find out more, please visit https://digital-library.theiet.org.

In collaboration with the United Nations and the International Publishers Association, the IET is a Signatory member of the SDG Publishers Compact. The Compact aims to accelerate progress to achieve the Sustainable Development Goals (SDGs) by 2030. Signatories aspire to develop sustainable practices and act as champions of the SDGs during the Decade of Action (2020–30), publishing books and journals that will help inform, develop, and inspire action in that direction.

In line with our sustainable goals, our UK printing partner has FSC accreditation, which is reducing our environmental impact to the planet. We use a print-on-demand model to further reduce our carbon footprint.

Published by The Institution of Engineering and Technology, London, United Kingdom

The Institution of Engineering and Technology (the "**Publisher**") is registered as a Charity in England & Wales (no. 211014) and Scotland (no. SC038698).

Copyright © The Institution of Engineering and Technology and its licensors 2025

First published 2024

All intellectual property rights (including copyright) in and to this publication are owned by the Publisher and/or its licensors. All such rights are hereby reserved by their owners and are protected under the Copyright, Designs and Patents Act 1988 ("**CDPA**"), the Berne Convention and the Universal Copyright Convention.

With the exception of:

(i) any use of the publication solely to the extent as permitted under:

   a. the CDPA (including fair dealing for the purposes of research, private study, criticism or review); or
   b. the terms of a licence granted by the Copyright Licensing Agency ("**CLA**") (only applicable where the publication is represented by the CLA); and/or

(ii) any use of those parts of the publication which are identified within this publication as being reproduced by the Publisher under a Creative Commons licence, Open Government Licence' or other open source licence (if any) in accordance with the terms of such licence, no part of this publication, including any article, illustration, trade mark, or other content whatsoever, may be used, reproduced, stored in a retrieval system, distributed, or transmitted in any form or by any means (including electronically) without the prior permission in writing of the Publisher and/or its licensors (as applicable).

The commission of any unauthorised activity may give rise to civil or criminal liability.

Please visit https://digital-library.theiet.org/copyrights-and-permissions for information regarding seeking permission to reuse material from this and/or other publications published by the Publisher. Enquiries relating to the use, including any distribution, of this publication (or any part thereof) should be sent to the Publisher at the address below:

The Institution of Engineering and Technology
Futures Place,
Kings Way, Stevenage
Herts, SG1 2UA,
United Kingdom

www.theiet.org

While the Publisher and/or its licensors believe that the information and guidance given in this publication is correct, an individual must rely upon their own skill and judgement when performing any action or omitting to perform any action as a result of any statement, opinion, or view expressed in the publication and neither the Publisher nor its licensors assume and hereby expressly disclaim any and all liability to anyone for any loss or damage caused by any action or omission of an action made in reliance on the publication and/or any error or omission in the publication, whether or not such an error or omission is the result of negligence or any other cause. Without limiting or otherwise affecting the generality of this statement and the disclaimer, while all URLs cited in the publication are correct at the time of press, the Publisher has no responsibility for the persistence or accuracy of URLs for external or third-party internet websites and does not guarantee that any content on such websites is, or will remain, accurate, or appropriate.

While every reasonable effort has been undertaken by the Publisher and its licensors to acknowledge copyright on material reproduced, if there has been an oversight, please contact the Publisher and we will endeavour to correct this upon a reprint.

Trademark notice: Product or corporate names referred to within this publication may be trademarks or registered trademarks and are used only for identification and explanation without intent to infringe.

Where an author and/or contributor is identified in this publication by name, such author and/or contributor asserts their moral right under the CPDA to be identified as the author and/or contributor of this work.

**British Library Cataloguing in Publication Data**

A catalogue record for this product is available from the British Library

**ISBN 978-1-83953-641-0 (hardback)**
**ISBN 978-1-83953-642-7 (PDF)**

Typeset in India by MPS Limited

*Cover Image:* Teera Konakan/Moment via *Getty Images*

# Contents

| | | |
|---|---|---|
| Preface | | xv |
| About the authors | | xix |

**Part I  Introduction to machine learning and neural networks** — 1

**1  Artificial intelligence, machine learning, and deep learning** — 3
  1.1  Artificial intelligence, machine learning, and deep learning: definition and relationship — 3
  1.2  What is machine learning? — 4
  1.3  Supervised learning — 5
    1.3.1  Classification — 6
    1.3.2  Regression — 6
  1.4  Unsupervised learning — 7
    1.4.1  Clustering — 7
    1.4.2  Dimensionality reduction — 9
  1.5  Reinforcement learning — 10
    1.5.1  Environment — 10
    1.5.2  Reward — 11
    1.5.3  Policy — 12
    1.5.4  Training of reinforcement learning algorithms — 13
  1.6  Summary — 13
  References — 14

**2  Deep neural networks** — 15
  2.1  What is an artificial neuron? — 15
  2.2  Neural network architectures — 16
  2.3  Choosing activation functions for multi-layer neural networks — 17
    2.3.1  Sigmoid function — 18
    2.3.2  Hyperbolic tangent function — 18
    2.3.3  Rectified linear unit — 19
    2.3.4  Softmax function — 21
  2.4  Learning algorithm — 21
    2.4.1  Perceptron learning rule — 21
    2.4.2  Adaptive linear neurons and the gradient descent learning method — 22
    2.4.3  Batch gradient descent — 24
    2.4.4  Dealing with large training datasets: stochastic gradient descent — 24

|   |   |   |   | |
|---|---|---|---|---|
|   | | 2.4.5 | A trade-off between gradient descent and SGD: Mini-batch gradient descent | 25 |
|   | | 2.4.6 | SGD with momentum: overcoming the noisy nature of SGD and mini-batch | 25 |
|   | | 2.4.7 | Adjusting the learning rate: adaptive gradient descent (AdaGrad) | 26 |
|   | 2.5 | Adjusting the deep neural network's coefficients through the error backpropagation | | 26 |
|   | | 2.5.1 | Multi-layer neural network | 27 |
|   | 2.6 | Overfitting and underfitting | | 30 |
|   | 2.7 | Summary | | 32 |
|   | References | | | 32 |

## Part II  Deep reinforcement learning    33

**3  Markov decision process**    35
- 3.1 A brief recap on reinforcement learning: a Markovian perspective    35
  - 3.1.1 Policy classification and expected reward    36
  - 3.1.2 Bellman's equation    37
  - 3.1.3 Agent vs. environment: learning the policy    38
- 3.2 Multi-arm bandit    39
  - 3.2.1 Exploration and exploitation    39
  - 3.2.2 Epsilon-greedy method    40
  - 3.2.3 Softmax selection    40
  - 3.2.4 Action value estimation approaches    41
- 3.3 Dynamic programming    43
- 3.4 Monte Carlo methods    46
  - 3.4.1 Estimation procedures    47
  - 3.4.2 General policy iteration in Monte Carlo methods    48
- 3.5 Temporal difference learning    49
  - 3.5.1 Estimation procedures    50
  - 3.5.2 General policy iteration in temporal difference learning: on-policy and off-policy    50
- 3.6 Summary    51
- References    51

**4  Value function approximation for continuous state-action space**    53
- 4.1 From tabular to function approximation method    53
- 4.2 Deep Q-learning    54
- 4.3 Methods for sampling the replay buffer    56
- 4.4 Double DQL    58
- 4.5 Dueling DQL    58
- 4.6 Comparison of the different networks    60
- 4.7 Summary    61
- References    62

## 5 Policy search methods for reinforcement learning — 63
5.1 The rationale behind policy search — 63
5.2 Policy gradient — 64
5.3 REINFORCE — 64
5.4 Natural policy gradient methods — 65
    5.4.1 Trust region policy optimisation — 67
    5.4.2 Proximal policy optimisation — 68
5.5 Deterministic policy gradient — 70
5.6 Summary — 71
References — 71

## 6 Actor-critic learning — 73
6.1 Brief introduction — 73
    6.1.1 Recap on policy gradient: actor-critic approach — 73
    6.1.2 Reducing variance with the advantage — 75
6.2 Different categories of methods — 76
    6.2.1 Deep deterministic polity gradient — 76
    6.2.2 Asynchronous advantage actor-critic (A3C) — 77
    6.2.3 Proximal policy optimisation — 77
6.3 Summary — 79
References — 79

## Part III Deep reinforcement learning in UAV-assisted 6G communication — 81

## 7 UAV-assisted 6G communications — 83
7.1 6G networks requirements — 83
    7.1.1 System efficiency — 85
    7.1.2 Guaranteed throughput — 85
7.2 Benefits of UAV integration — 85
    7.2.1 Main UAV-assisted services — 86
7.3 Open challenges — 87
    7.3.1 Efficient energy management — 87
    7.3.2 3D path planning — 88
7.4 Next research directions: digital twin and deep reinforcement learning — 88
    7.4.1 Resource management in UAV-enabled communications — 89
    7.4.2 Trajectory planning for UAV — 90
7.5 Summary — 90
References — 90

## 8 Distributed deep deterministic policy gradient for power allocation control in UAV-to-UAV-based communications — 97
8.1 Introduction — 97
8.2 System model and problem formulation — 100

| | | |
|---|---|---|
| 8.3 | Multi-agent power allocation problem in U2U-based communications: distributed deep deterministic policy gradient approach | 103 |
| | 8.3.1 Value function | 103 |
| | 8.3.2 Policy search | 104 |
| | 8.3.3 Distributed deep deterministic policy gradient | 105 |
| 8.4 | Sharing deep deterministic policy gradient for multi-agent power allocation problem in D2D-based V2V communications | 106 |
| 8.5 | Simulation results | 110 |
| 8.6 | Summary | 115 |
| References | | 115 |

## 9 Non-cooperative energy-efficient power allocation game in UAV-to-UAV communication: a multi-agent deep reinforcement learning approach — 119

| | | |
|---|---|---|
| 9.1 | Introduction | 119 |
| 9.2 | Related work | 121 |
| 9.3 | System model and problem formulation | 123 |
| 9.4 | Reinforcement learning for energy-efficient power allocation game in U2U communication | 124 |
| | 9.4.1 Single-agent Q-learning | 124 |
| | 9.4.2 Multi-agent Q-learning approach | 125 |
| 9.5 | Deep reinforcement learning for power allocation optimisation in U2U communication | 127 |
| | 9.5.1 Deep Q-learning | 127 |
| | 9.5.2 Double deep Q-learning | 128 |
| | 9.5.3 Dueling deep Q-learning | 128 |
| 9.6 | Simulation results | 131 |
| | 9.6.1 Performance comparison | 132 |
| | 9.6.2 Scalability analysis | 132 |
| | 9.6.3 Exploration/exploitation analysis | 132 |
| | 9.6.4 Running time analysis | 134 |
| 9.7 | Summary | 135 |
| References | | 135 |

## 10 Real-time energy harvesting-aided scheduling in UAV-assisted D2D networks — 139

| | | |
|---|---|---|
| 10.1 | Introduction | 139 |
| | 10.1.1 State of the art and challenges | 140 |
| | 10.1.2 Contributions | 142 |
| 10.2 | System model and problem formulation | 142 |
| 10.3 | Preliminaries | 146 |
| | 10.3.1 Value function | 146 |
| | 10.3.2 Policy search | 147 |

Contents    xi

|  | 10.4 | Energy harvesting time scheduling in UAV-powered D2D communications: a deep deterministic policy gradient approach | 148 |
|---|---|---|---|
|  | 10.5 | Efficient learning with proximal policy optimisation algorithms to solve the energy harvesting time scheduling problem in D2D communications assisted by UAV | 150 |
|  |  | 10.5.1 Clipping surrogate method | 150 |
|  |  | 10.5.2 Kullback–Leibler divergence penalty | 150 |
|  | 10.6 | Simulation results | 151 |
|  |  | 10.6.1 Performance comparison | 151 |
|  |  | 10.6.2 Parameter analysis | 153 |
|  |  | 10.6.3 Computational complexity | 156 |
|  | 10.7 | Summary | 156 |
|  | References |  | 157 |

**11 3D trajectory design and data collection in UAV-assisted networks**    **161**

|  | 11.1 | Introduction | 161 |
|---|---|---|---|
|  |  | 11.1.1 Related contributions | 162 |
|  |  | 11.1.2 Contributions and organisation | 163 |
|  | 11.2 | System model and problem formulation | 165 |
|  |  | 11.2.1 Observation model | 165 |
|  |  | 11.2.2 Game formulation | 166 |
|  | 11.3 | Preliminaries | 169 |
|  | 11.4 | An effective deep reinforcement learning approach for UAV-assisted IoT networks | 170 |
|  | 11.5 | Deep reinforcement learning approach for UAV-assisted IoT networks: A dueling deep Q-learning approach | 172 |
|  | 11.6 | Simulation results | 174 |
|  |  | 11.6.1 Expected reward | 174 |
|  |  | 11.6.2 Throughput comparison | 178 |
|  |  | 11.6.3 Parametric study | 181 |
|  | 11.7 | Summary | 186 |
|  | References |  | 186 |

**Part IV  Deep reinforcement learning in reconfigurable intelligent surface-empowered 6G communications**    **191**

**12 RIS-assisted 6G communications**    **193**

|  | 12.1 | Introduction | 193 |
|---|---|---|---|
|  | 12.2 | RIS technology: a brief overview | 194 |
|  | 12.3 | RIS in next-generation wireless networks | 195 |
|  |  | 12.3.1 RIS-assisted multicell networks | 196 |
|  |  | 12.3.2 RIS-aided non-orthogonal multiple access | 196 |
|  |  | 12.3.3 RIS for simultaneous wireless information and power transfer | 196 |
|  |  | 12.3.4 RIS-assisted mobile edge computing networks | 197 |

|  |  |  |
|---|---|---|
| | 12.3.5 RIS for physical layer security | 197 |
| 12.4 | RIS-empowered UAV-assisted communications | 198 |
| | 12.4.1 Enhancing air-to-ground (A2G) links | 198 |
| | 12.4.2 RIS-equipped UAV communications | 198 |
| 12.5 | Challenges and research directions ahead | 199 |
| | 12.5.1 Channel state estimation techniques | 199 |
| | 12.5.2 RIS deployment strategies | 200 |
| | 12.5.3 Mobility management | 200 |
| | 12.5.4 The use of AI in RIS-assisted communications | 200 |
| 12.6 | Summary | 201 |
| References | | 201 |

## 13 Real-time optimisation in RIS-assisted D2D communications    203
13.1 Introduction    203
13.2 System model and problem formulation    205
13.3 Joint optimisation of power allocation and phase shift matrix    207
13.4 Simulation results    209
13.5 Summary    214
References    214

## 14 RIS-assisted UAV communications for IoT with wireless power transfer using deep reinforcement learning    217
14.1 Introduction    217
    14.1.1 State of the art    218
    14.1.2 Contributions    220
    14.1.3 Organisation    221
14.2 System model and problem formulation    221
    14.2.1 Channel model    222
    14.2.2 Power transfer phase    223
    14.2.3 Information transmission phase    224
14.3 Hovering UAV for downlink power transfer and uplink information transmission in RIS-assisted UAV communications    225
    14.3.1 Preliminaries    225
    14.3.2 DDPG method    226
    14.3.3 Game solving    227
14.4 Joint trajectory, transmission power, energy harvesting time scheduling, and the RIS phase shift optimisation using deep reinforcement learning    228
14.5 Proximal policy optimisation technique for joint trajectory, power allocation, energy harvesting time, and the phase shift optimisation    229
14.6 Simulation results    231
14.7 Summary    237
References    237

| | | |
|---|---|---|
| **15 Multi-agent learning in networks supported by RIS and multi-UAVs** | | **241** |
| 15.1 | Introduction | 241 |
| | 15.1.1 Related work | 242 |
| | 15.1.2 Contributions | 243 |
| 15.2 | System model and problem formulation | 244 |
| | 15.2.1 System model | 244 |
| | 15.2.2 Problem formulation | 246 |
| 15.3 | Preliminaries | 247 |
| | 15.3.1 Value function | 247 |
| | 15.3.2 Policy search | 248 |
| 15.4 | Centralised optimisation for power allocation and phase-shift matrix | 248 |
| 15.5 | Parallel DRL for joint power allocation and phase-shift matrix Optimisation | 251 |
| 15.6 | Proximal policy optimisation for centralised and decentralised problem. | 251 |
| 15.7 | Simulation results | 253 |
| 15.8 | Summary | 259 |
| References | | 259 |
| **Index** | | **263** |

# Preface

Today, we are experiencing an unprecedented surge in smart wearable electronic devices like smartphones and tablets, alongside the rise of the Internet-of-Things (IoT) paradigm. This interconnected ecosystem enables devices to work together, offering innovative services such as smart manufacturing and autonomous driving, known as Industrial IoT and Internet of Vehicles (IoV). This rapidly evolving trend is driving the creation of highly complex communication scenarios. Current communication technologies, such as conventional terrestrial base stations, may struggle to meet the demanding Key Performance Indicators (KPIs) for the delivery of advanced disruptive services and applications like extended reality (ER), tactile internet, intelligent transportation systems, global ubiquitous connectivity, and pervasive intelligence. Then, as already happened with the previous generation of telecommunication systems, these concerns have led both industry and academia to start extensive discussions and research activities for the deployment of 6G wireless communication technology.

Compared to the current 5G networks that are still under deployment, the upcoming 6G communication technologies are expected to provide considerable improvements and a higher level of communication performance. More specifically, it is envisaged that 6G networks will be able to provide connection density and data rates in the order of $10^7$ users/km$^2$ and 1 Terabits-per-second (Tbps), respectively. In addition, end-to-end communication latency less than 10 μs, spectral efficiency up to 90 bps/Hz in downlink and 45 bps/Hz in uplink, as well as connection reliability of 99.99999% are also expected to be achieved. The achievement of these stringent performances is expected to be reached thanks to the introduction of innovative communication technologies that 6G are expected to introduce at the physical layer of the network. Among these, the use of both terahertz (THz) and sub-terahertz frequency bands for signal transmission will play an important role. First of all, the use of these frequency bands is driven by the problem of spectrum scarcity, which we are experiencing at lower frequencies, including the millimetre waves. This will allow for a higher number of users to be connected. Furthermore, in light of substantial advancements in electronics and the large bandwidth availability, the use of the THz spectrum will enable wireless communication links with Tbps capacity. However, although the usage of THz bandwidth holds the potential to provide unprecedented wireless communication capabilities, its adoption results are limited by the fact that THz communications signals will be subject to strong penetration losses.

The adoption of unmanned aerial vehicles (UAVs) and reflective intelligent surfaces (RISs) has been recognised as a promising solution to provide a better propagation environment and overcome the problem of signal attenuation in the THz bandwidth. On the one hand, compared to traditional terrestrial base stations, using UAVs to provide signal coverage represents a more flexible and cost-effective solution, as they are easy to deploy and offer the possibility of establishing line-of-sight (LoS) links with ground users, thereby improving signal quality. On the other hand, the use of RISs enables control over the propagation environment. RISs are essentially arrays of micro or nano-scale structures that, based on the wavelength of the incident wireless signal, exhibit different macroscopic responses such as reflection towards a specific direction. Thus, UAVs and RISs hold the potential to fully promote a controllable propagation environment necessary for the deployment of next-generation 6G networks. However, several obstacles hinder the deployment and full exploitation of UAV and RIS-assisted network capabilities, mainly related to the need for frequent resource optimisation to meet network requirements. Currently, most frameworks for optimal resource allocation in wireless networks rely on conventional convex optimisation techniques, which only work well for small- and medium-scale optimisation problems. Due to the complex nature of UAV-based and RIS-assisted communication scenarios, these techniques often fall short of meeting requirements such as near-zero end-to-end communication latency. To this end, the adoption of artificial intelligence (AI), specifically deep reinforcement learning (DRL)-based algorithms, represents the most promising approach to solve complex resource optimisation problems in wireless networks, ensuring the achievement of network requirements.

The book chapters are arranged into four parts according to their topics. More specifically, the first two parts provide background and fundamentals on artificial intelligence and deep reinforcement learning, while the last two illustrate how these principles can be applied to solve complex optimisation problems in the context of UAV-enabled and RIS-assisted networks. In **Part I**, there are two chapters that provide an essential introduction to artificial intelligence and deep neural networks. Specifically, *Chapter 1* illustrates the relationship between artificial Intelligence, machine learning, and deep learning. After introducing these concepts, *Chapter 2* delves into the principles of deep reinforcement learning.

The basis provided in **Part I** will be very helpful for readers to understand the contents of **Part II**, which contains four chapters illustrating the most relevant aspects of deep reinforcement learning. In particular, this part starts with a brief introduction to Markov decision processes in *Chapter 3*. Subsequently, it delves deeper into the concepts of value function approximation and policy search methods, illustrated in *Chapters 4 and 5*, respectively. Finally, it concludes with *Chapter 6*, which provides an introduction to actor-critic learning.

**Part III** consists of five chapters focused on covering the aspects of UAV-enabled networks. This part starts with *Chapter 7*, which provides an overview of the potentialities and benefits of implementing UAV-enabled networks and also illustrates the main challenges and urgent research directions in this area. *Chapter 8* demonstrates how the use of distributed deep deterministic policy gradient can solve the problem of power control in UAV-to-UAV communication, while *Chapter 9* explores the potential of multi-agent deep reinforcement learning in solving the energy efficiency optimisation problem in UAV-to-UAV communications. Subsequently, *Chapter 10* illustrates how the problem of real-time energy harvesting and communication scheduling in UAV-enabled networks can be addressed using a deep deterministic policy gradient method. Finally, *Chapter 11* shows how the problem of 3D trajectory design and data collection in UAV-assisted networks can be efficiently solved through a deep reinforcement learning approach.

On the other hand, **Part IV** focuses on illustrating how deep reinforcement learning can be efficiently used to solve optimisation problems in RIS-assisted networks. An overview of RIS-assisted 6G communications is provided in *Chapter 12*, highlighting the main advantages and related challenges of integrating RIS surfaces into next-generation networks. Following this introductory part, the most relevant optimisation problems in RIS-assisted networks are addressed. Specifically, *Chapter 13* illustrates a deep reinforcement learning approach for solving the joint optimisation problem of power allocation and phase-shift matrix for RIS-assisted D2D communications. *Chapter 14* deals with joint resource allocation in RIS-assisted UAV communications for IoT with wireless power transfer using deep reinforcement learning. Finally, *Chapter 15* explains how multi-agent learning can be beneficial for solving the joint power allocation and phase-shift matrix optimisation in networks supported by RIS and multi-UAVs.

The creation of this book has been made possible through the invaluable assistance, support, and experience of many people, to whom the authors wish to express their heartfelt gratitude. The authors also want to extend their immense appreciation to the diligent reviewers whose input greatly enhanced the quality of each chapter. Finally, the authors would like to convey their deep appreciation for the unwavering support provided by the IET staff during the preparation process of this book.

**Antonino Masaracchia, Khoi Khac Nguyen,
Trung Q. Duong, and Vishal Sharma
Authors**

# About the authors

**Antonino Masaracchia** is a Lecturer at Queen Mary University of London. Previously, he was a research fellow at the Centre for Wireless Innovation at Queen's University Belfast, UK. His research activities are focused on optimal resource allocation for enabling URLLCs services in future 5G and 6G networks, and the investigation of innovative real-time optimisation systems for future wireless networks with digital twins. His research interests include digital twins, URLLCs, UAV communications, real-time optimisation, and AI and ML techniques for wireless communication networks. He has authored or co-authored 50+ books, book chapters, journals, and conference papers. He has been awarded with the seal of excellence for the project proposal UAV-SURE and UAV-DRESS submitted under the Horizon 2020 s Marie Sklodowska-Curie actions call H2020-MSCA-IF-2020 (September 2020), and Horizon Europe Marie Skodowska-Curie Actions call HORIZON-MSCA-2021-PF-01-01 MSCA Postdoctoral Fellowships (October 2021), respectively.

**Khoi Khac Nguyen** completed his PhD degree with the School of Electronics, Electrical Engineering and Computer Science, Queen's University Belfast, UK. His research interests include machine learning and deep reinforcement learning for real-time optimisation in wireless networks, reconfigurable intelligent surfaces, unmanned air vehicle (UAV) communication, and massive Internet of Things (IoTs).

**Trung Q. Duong** is a Canada Excellence Research Chair (CERC) and full professor at Memorial University, Canada. He is also an adjunct chair professor in telecommunications at Queen's University Belfast, UK and a research chair of the Royal Academy of Engineering. His research interests include machine learning, real-time optimisation, data analytics, and 5G–6G networks. He has authored or co-authored 500 + books, book chapters, journals, and conference papers. He is a recipient of the prestigious Royal Academy of Engineering Research Fellowship (2016–20) and won the Newton Prize in 2017.

**Vishal Sharma** is a senior lecturer in the School of Electronics, Electrical Engineering and Computer Science (EEECS) at Queen's University Belfast (QUB), Northern Ireland, UK. At QUB, he focuses on cyber defence and security and leads research on drone security, digital twins (DT), and blockchain systems. Since 2022, he has been affiliated with the Global Innovation Institute as a fellow and works with the Centre for Secure Information Technologies (CSIT) and the Centre for Intelligent Sustainable Computing (CISC). He is the director of the Innovation-by-design lab at QUB. He leads the British Computer Society (BCS) Accreditation for QUB and is the chair of the Computer Science Programme Review Working Group. He is the co-investigator for the Northern Ireland Advanced Research and Engineering

Centre (ARC). He has authored/co-authored more than 140 journal/conference articles and book chapters, co-edited four books and won seven best paper awards. He has served on the editorial board of *IEEE Communications Magazine* and as section editor-in-chief of Drones journal. Currently, he serves as associate editor of CAAI Transactions on Intelligence Technology, IET Networks, and ICT Express. He is also the co-chair of the IEEE UK and Ireland Diversity, Equity, and Inclusion Committee. He is a senior member of IEEE and a professional member of ACM. He received his PhD degree in Computer Science and Engineering from Thapar University, India.

*Part I*

# Introduction to machine learning and neural networks

*Chapter 1*
# Artificial intelligence, machine learning, and deep learning

In the modern era, we are witnessing a revolutionary transformational process where the exponential diffusion of smart and connected electronic devices fosters the emergence of disruptive services, such as Internet-of-Things (IoT), smart manufacturing, autonomous driving, and virtual/augmented reality (AR/VR). However, to fully unleash the potential of these technologies, it is necessary to address challenges related to the vast amount of data generated by these devices. Additionally, optimising network resources is crucial to guarantee very low communication delays and error-free communications. This chapter provides a brief introduction to the concepts of artificial intelligence (AI), machine learning (ML), and deep learning (DL), with a main focus on ML-related aspects, while DL-related concepts will be covered in later chapters. These concepts form the basis for understanding the contents provided within this book. In particular, this chapter aims to provide the reader with a clear understanding of what AI, ML, and DL are, how they work, and their pivotal role in the deployment of next-generation networks.

## 1.1 Artificial intelligence, machine learning, and deep learning: definition and relationship

AI has been conceived in response to creating computer systems capable of performing tasks that typically require human intelligence. Some include learning from data, understanding natural language, speech recognition, visual perception, and problem-solving. In other words, AI is intended to create machines able to replicate or simulate human-like cognitive abilities, such as learning from experience, making decisions, and adapting to new situations. The birth of AI was conventionally placed within the first half of the twentieth century. In 1950, Alan Turing proposed the Turing Test as a measure of a machine's ability to exhibit human-like intelligence. Subsequently, in 1956, the term AI was fully coined at the Dartmouth conference, where the attendees, including John McCarthy, Marvin Minsky, and others, envisioned the creation of machines that could simulate human intelligence. Despite this great enthusiasm, the field of AI experienced a period known as "AI winter" in the 1980s, marked by reduced funding and diminished interest due to unmet expectations.

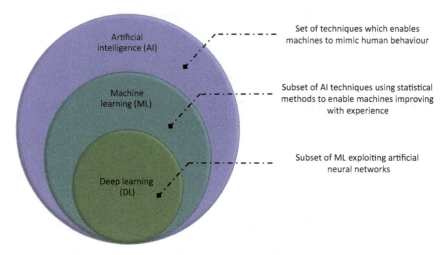

*Figure 1.1  Relationship between AI, ML, and DL*

The winter period persisted for nearly two decades until the late twentieth century. Influenced by the adoption of statistical approaches, the availability of vast amounts of data, and increased computing power, interest in AI was rekindled. This resurgence has not only gained attention but continues to grow in both industry and academia.

Within its definition, AI encompasses a broad range of techniques and theories aimed at embedding human intelligence into computational machines. In this book, we will focus on a sub-field of AI, which is ML, and specifically within the context of DL techniques. Let us attempt to clarify the relationship between these fields. As illustrated in Figure 1.1, the relationship between these fields is intricately connected. As mentioned earlier, AI is the fundamental concept at the core of creating intelligent agents capable of mimicking human-like cognitive functions. Within this field, ML represents a subset of AI, which is mainly focused on the development of statistical-based models and algorithms. Such algorithms are designed to carry out a specific task by learning from data and improving their performances over time. Finally, DL is a specialised form of ML that employs artificial neural networks to model and process complex patterns and representations. DL was mainly used to perform image and speech recognition tasks. However, nowadays, DL-based techniques have a very broad range of applications. The content and applications provided within this book are mainly focused on DL-based techniques. However, to provide all the essential insights and basics to those not familiar with this field, the subsequent part of this chapter will provide a brief overview of ML-related aspects.

## 1.2  What is machine learning?

The early development of intelligent applications was perceived as the design of software codes mainly consisting of if–else rules to execute when particular conditions

*Artificial intelligence, machine learning, and deep learning* 5

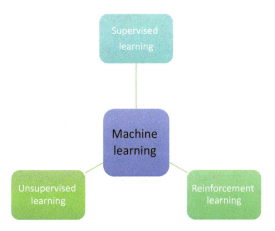

*Figure 1.2 Different types of ML techniques*

occurred during the execution of the code itself or when a specific input was provided by the final user. This process is completely useful and efficient when the code designer clearly understands the system that needs to be modelled. However, this approach often falls short due to the following reasons:

- The design of such rules requires a human expert with a deep understanding of how a decision should be made,
- The reusability of the code results is very limited since changing the decisional policy might require rewriting the code.

Adopting ML-based approaches represents an optimal approach to overcome these issues. Indeed, ML is a branch of AI that explores the learning theory of pattern recognition from data, empowering computers to learn and make decisions either with minimal or without explicit human intervention. At its essence, ML uses statistical methods to learn from data samples, enabling the realisation of intelligent systems that can recognise patterns within vast datasets and improve their performance over time. In other words, the key aspect of ML-based algorithms lies in their ability to adapt themselves autonomously and refine their inner models as they encounter new information. For these reasons, in recent years, ML-based methods have become popular in different technological areas, ranging from image recognition to automatic recommendations of which movies to watch to what food to order or which products to buy. The automatic process of classifying data and the self-adaptability of ML algorithms encompasses various approaches, as illustrated in Figure 1.2. These include supervised, unsupervised, and reinforcement learning, which will be briefly illustrated in the following subsections.

## 1.3 Supervised learning

Supervised learning is a class of ML techniques that infer how the model learns from labelled data and uses the trained model to predict the unlabelled data. For this reason, it represents one of the most commonly used and successful types of ML techniques. Simply put, a supervised ML technique can be represented as a black-box which aims at building a model that represents the relationship between the input data and the correct output label. In doing so, it is important to mention that supervised learning is used whenever we have examples of input/output pairs that basically will represent our training set. Once trained on this dataset, the ML will be used to accurately predict the unseen data. As one can note, the selected training set must satisfy some technical requirements. In particular, it needs to (i) guarantee a good level of randomness, and (ii) cover as many general cases as possible. This book is not intended to delve into the details of data preprocessing and feature extraction procedures. Readers interested in understanding how training datasets influence the performance of supervised ML models can refer to comprehensive studies presented in [1]. The following subsections illustrate the two main categories of supervised learning: **classification** and **regression**.

## 1.3.1 Classification

The classification concept probably represents the most intuitive way to understand how supervised learning works. Imagine, for example, having a generic set of pictures representing animals. Such pictures have different features, such as image resolution, number of pixels and colour associated with each pixel. At the same time, each picture has a label representing the animal type. Based on these assumptions, you want to build an ML-based algorithm that, once trained on that specified dataset, will be able to classify, i.e. assign the correct label to future unseen pictures. This is what exactly supervised ML algorithms do: based on the features of the data, the model can label the data into a category. In ML classification, the target label assumes a categorical nature, and the input data is typically expressed as a feature vector encapsulating various aspects of each sample. This representation facilitates the algorithm's ability to discern patterns and make informed predictions. Some popular algorithms employed for classification tasks include logistic regression, adept at modelling the probability of a sample belonging to a particular class; decision trees, which navigate hierarchical decision structures; and $k$-nearest neighbours, which classify based on the prevailing label among nearby data points. The applications of classification are diverse and impactful. In the context of human recognition systems, some ML algorithms can identify individuals based on distinctive features. On the other hand, in the intricate landscape of biological data analysis, ML classification can be helpful in categorising genetic sequences, biomarkers, or other biological entities. For this reason, the ability of classification algorithms to extract meaningful insights from complex datasets has and continues to obtain attention across a myriad of domains.

## 1.3.2 Regression

In contrast to classification, which focuses on assigning discrete labels to data, a regression task entails predicting a continuous output variable within a range of values based on a given set of features. Analogous to classification, the training dataset for regression comprises samples with associated features, each paired with a continuous label. Regression models find broad application across various domains, significantly contributing to finance, forecasting, and estimation. For instance, envision developing an ML regression model tailored to forecast the price of a house. This model might consider an array of features, including the property's location, the number of rooms, the types of building materials used, and the level of thermal insulation. By analysing these features, the regression model can provide a nuanced and continuous prediction of the house's price, allowing for a more granular understanding of the real estate market. Another illustrative example lies in estimating an individual's salary using regression. Factors such as age, educational background, and prior professional experiences can be incorporated into the model to predict a continuous output, namely, the expected salary. This application showcases the versatility of regression in handling real-world scenarios where the outcome is not confined to discrete categories.

## 1.4 Unsupervised learning

Unsupervised learning is the second family of ML algorithms. As the definition suggests, these algorithms are intended to work on unlabelled datasets. Indeed, in contrast to supervised ML, where the final goal is to make predictions, unsupervised ML algorithms are designed with the main objective of feature extraction from data, i.e. finding important relationships among the samples provided within the training data. In the following subsections, we will discuss the most common unsupervised ML tasks, namely, **clustering** and **dimensionality reduction**, which are generally used for anomaly detection and data visualisation.

### 1.4.1 Clustering

Clustering tasks are mainly exploited to divide the original dataset into distinct groups known as clusters. Once the clusters are defined, points within the same cluster exhibit similar features or characteristics, while those in different clusters demonstrate dissimilarity. In this way, similar to supervised learning algorithms, it is possible to say that once trained, the clustering algorithms aim at assigning a label to a specific data point. Among the most popular clustering algorithms in the literature, K-means, hierarchical, and density-based clustering are the most commonly used. To give an idea of how these algorithms work, let us briefly introduce the K-means clustering algorithm, which represents the simplest one. Indeed, as per its name, it tries to find the cluster centres representative of certain regions of the data. In doing so, it alternates between different steps (see Figure 1.3). First, it assigns each data point to the closest cluster centre, called the centroid. At the beginning, the number of centroids $K$ is provided and the centroid is selected randomly. Once the grouping based on minimum distance is calculated, the new centroids are calculated as the average of all the points within the cluster, and the distance from each new centroid is calculated again. If there are points to move, the

8  *DRL for RIS and UAV Empowered Smart 6G Communications*

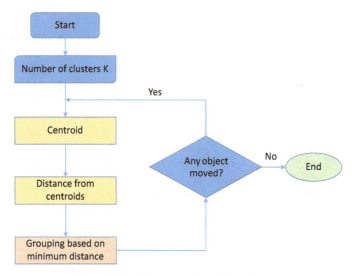

*Figure 1.3  K-means algorithm diagram*

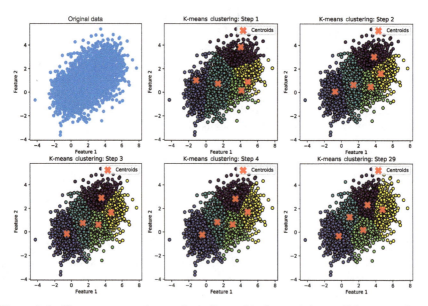

*Figure 1.4  K-means operation on data. From this figure, it is possible to note how the centroids are moved step by step. Figure generated using tools and datasets from Scikit-learn library [2].*

process is repeated. Otherwise, it is terminated. An example of how the K-means algorithm operates on data is provided in Figure 1.4, where the original data (top-left) is successfully clustered into five clusters after 29 iterations of the algorithm (bottom-right).

## 1.4.2 Dimensionality reduction

In some datasets, the features are not always useful and effective in learning. The noise in the features can make the model biased and predict inefficiently. Thus, dimension reduction methods are often used for data preprocessing before putting it into the other learning model. A number of useful features can be obtained, while the others will be discarded. It makes learning faster when the input is not complex and more efficient when the data is more balanced. Principal component analysis (PCA) is the simplest and most widely used among the most popular algorithms for dimensionality reduction. PCA identifies the directions (principal components) in which the data varies the most. It then projects the data into a new space defined by these principal components. It is usually used in image compression, which means reducing the dimensions of an image while preserving important features. For example, if you have a high-resolution image, PCA can help represent it with fewer components. This is illustrated in Figure 1.5, which contains the first 15 of the 100 principal components of the labelled faces in the Wild dataset [2]. This means that the generic image for that dataset can be expressed as a linear combination of each component with appropriate coefficients. Let **IM** be the set of all pictures in the dataset, with **C**

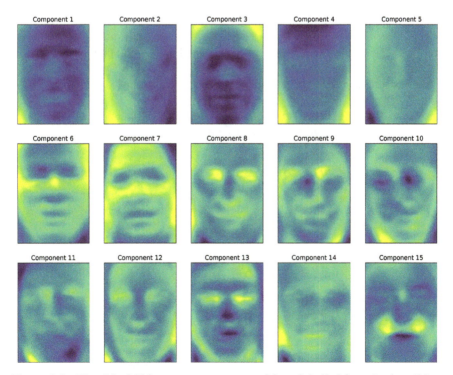

*Figure 1.5* First 15 of 100 components extracted from labelled faces in the wild dataset. Figure generated using tools and datasets from Scikit-learn library [2].

being the components, and $w^j$ being the vector of weights for the $j$-th picture $\in \mathbf{IM}$. We can reconstruct the generic $IM^j$ picture as follows:

$$IM^j = \sum_{i=1}^{100} w^i \cdot c_i, \quad \forall IM^j \in \mathbf{IM}, c_i \in C. \tag{1.1}$$

Another example can be found in the finance market, where the PCA is used to analyse the correlations between different stocks and then identify the principal components representing the most significant movements in stock prices.

## 1.5 Reinforcement learning

Reinforcement learning (RL) is another sub-class of machine learning. In contrast to supervised and unsupervised ML learning algorithms that operate on static datasets, RL operates on dynamic data constantly coming from a dynamic environment. Another distinctive feature of RL-based algorithms is that they are not intended to predict a value or assign labels. Instead, they are intended to find the best sequence of actions that will generate the optimal outcome. As illustrated in Figure 1.6, RL procedures are represented by an agent software. The agent observes the data coming from the environment, referred to as a state. Based on this state and its policy, the action is decided. Once the action is applied to the environment, it changes state and produces a reward to the agent for that action. Using this new information, the agent can determine whether that action was good and should be repeated, or if it was bad and should be avoided. The observation-action-reward cycle, often referred to as trial-and-error learning, continues until the optimal policy, i.e. mapping state to actions that maximise the long-term reward, is obtained. Since RL, more specifically, deep RL (DRL), represents a fundamental part of this book, the rest of this section will provide a brief overview of the main component blocks.

### 1.5.1 Environment

Essentially, the environment represents the space where the RL-based algorithm will be placed. The agent will take observations from the environment and send actions

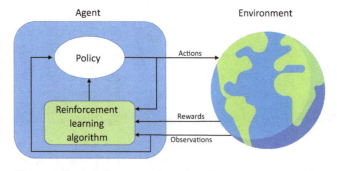

*Figure 1.6 Reinforcement learning diagram representation*

into it; from each action sent, the agent will receive a reward. One of the main reasons why RL is so powerful and still continues to receive attention is that the agent can still learn how to interact with it, even if it is completely unaware of its dynamics. For this reason, RL algorithms are often referred to as **model-free learning algorithms**. Indeed, it may be possible to put a model-free RL agent into any system and let the agent learn the optimal policy. It only needs access to the observations (states) and rewards from the environment itself. However, setting a model-free RL model can have some drawbacks. The most intuitive drawback is that this approach can require considerable time before the agent learns the optimal policy. This is because the agent will need to explore low-reward areas during the learning process. But sometimes, the designer knows some parts of the environment that are not worth exploring, and embedding this information in the RL model will be definitively helpful in reducing the time necessary for the agent to find the optimal policy. This approach is referred to as **model-based** RL and is proven to lower the time the agent needs to find the optimal policy by guiding the agent away from areas of the state space known to have low rewards.

### 1.5.2 Reward

Once the environment is set and all the necessary information representing the environment state is defined, the next step is to design the set of possible actions that the agent can perform and what and how it will be rewarded after performing such actions. This, in essence, means designing a reward function so that the learning algorithm can perceive when the policy is improving and how to improve it to achieve the highest long-term reward. Basically, the reward can be thought of as a function defined as follows:

$$\text{reward} = \text{function}(\text{action}, \text{state}). \tag{1.2}$$

More precisely, the definition provided in (1.2) represents a function that produces a scalar value representing the rightness of the agent being in a particular state and taking a particular action. There are no constraints on how to design a reward function. However, even in the environment, embedding some domain-specific information into the reward function is a good practice. Although it might sound easy to do, crafting a good reward function represents one of the more difficult tasks in RL.

The design of a reward function also comes with some issues. The first one is the case of **sparse rewards**. This happens when the final goal for which we want to obtain the optimal policy comes after a long sequence of actions. Imagine, for example, the case of an autonomous driving vehicle for which we want to find the optimal policy allowing it to drive for a certain distance without causing accidents. Then, the reward will be provided to the agent only when the destination is reached, respecting the accident-free requirements. The challenge associated with sparse rewards lies in the possibility of the agent meandering for extended durations, experimenting with various actions and exploring numerous states without obtaining any rewards. Consequently, the lack of rewards impedes the learning process, as the agent gains little to no valuable insights during its exploration.

Another issue to take into consideration is the **exploration vs. exploitation** trade-off. This comes from the fact that in RL, an agent interacts with an environment and the choices it makes determine the information it receives and, therefore, the information from which it can learn. However, there is always the chance that some areas of the environment that are not explored can provide higher rewards. In this case, it is legit to wonder if the agent should exploit the environment by choosing the actions that collect the most rewards that it already knows about or choosing actions that explore parts of the environment that are still unknown. In this way, finding a balance between exploration and exploitation is necessary since the former might take a long time to converge. In contrast, the latter might converge into a sub-optimal 'greedy' policy.

## 1.5.3 Policy

A policy can be viewed as a function that takes in state observations and outputs actions. One can find two different procedures to structure such type of mapping:

- Direct approach, also known as policy-mapping, where there is a direct mapping between state and actions to take;
- Indirect approach consisting of looking at other metrics values to infer the optimal mapping. This is also referred to as value-based mapping.

These two approaches will be covered in depth within the next chapters of this book. However, to give an introductory idea, we briefly illustrate the case of representing a policy through a table. As shown in Figure 1.7, it is possible to create a policy using a table where the rows represent the states and the columns represent all the

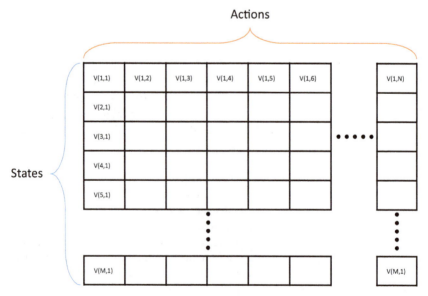

Figure 1.7  *Q-table with M states and N possible actions*

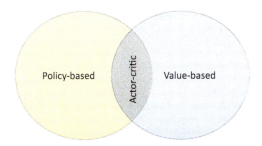

*Figure 1.8 Relationship between different RL training methods*

possible actions. The value inside each state-action entry represents the goodness of taking that particular action when in that specific state. This type of table is called a Q-table. Then, training an agent with a Q-table would consist of determining the correct values for each state-action pair in the table. As one can notice, this type of approach is feasible only when the state-action space has a finite number of elements that can be represented by a table. A common approach to deal with continuous state-action spaces is defining a function that, for each state, provides an action. This, as we will see later, is usually done through the adoption of neural networks.

### 1.5.4 Training of reinforcement learning algorithms

Once the environment, rewards, and policy structures are fully defined, the final step is to proceed with training the RL model. In mathematical terms, training an RL model corresponds to finding the optimal set of the parameters of the policy, which maximises the long-term reward. Indicating with $F_\mu^P(\mathscr{S})$ the policy function with parameters $\mu$, with $\mathscr{S}$ the state space, and with $R(\mathscr{S}, \mathscr{A})$ the reward of being in the state $\mathscr{S}$ and taking an action $\mathscr{A}$, we can express the training as the following optimisation problem:

$$\max_{\mu} \quad R(\mathscr{S}, \mathscr{F}_\mu^\mathscr{P}(\mathscr{S})). \tag{1.3}$$

There are different types of training procedures for RL-based algorithms that can broadly divided into: (i) value-based, (ii) policy-based, and (iii) actor-critic, with their relationship depicted in Figure 1.8. This will be covered in detail later since it is mainly based on the concept of using neural networks.

## 1.6 Summary

This chapter has first provided a brief introduction to AI, ML, and DL concepts, illustrating on what they consist of and their relationship. Subsequently, it has focused on the main area of ML. It discussed the main three classes, which are supervised learning, unsupervised learning, and reinforcement learning, representing the essential concepts to understand the contents provided within this book. The next chapter of this book will provide an overview of neural networks, which are the building blocks towards realising DL-based algorithms.

## References

[1] Müller AC. *Introduction to Machine Learning with Python*. O'Reilly Media, Inc.; 2016.
[2] Pedregosa F, Varoquaux G, Gramfort A, *et al.* Scikit-learn: Machine Learning in Python. *Journal of Machine Learning Research*. 2011;12:2825–2830.

# Chapter 2
# Deep neural networks

Within this chapter, the reader will be provided with the main understanding of neural network structure, starting from the basic definition of neurons to the organisation of neurons into structures forming a proper network. Once these concepts are explained, the most popular methods for training such neural networks will be briefly introduced and illustrated. In essence, this chapter represents the foundation for understanding the concepts of DL- and DRL-based algorithms.

## 2.1 What is an artificial neuron?

Artificial neurons represent a way to mimic how the human brain works. Its first appearance can be traced back to the mid-twentieth century when the so-called McCulloch–Pitt (MCP) neuron model was published. Starting from a biological structure, neurons are interconnected and in charge of transmitting information throughout the brain via electric signals. According to the MCP model (see Figure 2.1), a neuron can be represented as a gate that (i) receives multiple electric signals as inputs through the dendrites, (ii) integrates the inputs through the cell nucleus, and (iii) if the accumulated signal exceeds a certain threshold, an output electric signal is provided through the axon terminals. To this biological model, it is possible to build a mathematical model based on the context of binary classification. For example, let $\mathbf{x} \in \mathbb{R}^M$ be an $m$-dimensional input, i.e. an input with $M$ features arriving at the dendrites, with $\mathbf{w} \in \mathbb{R}^M$ being an $m$-dimensional vector of weights,

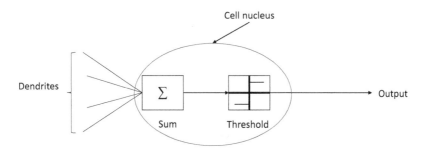

*Figure 2.1   Model of a biological neuron*

and $\phi(z)$ a decision function. We can describe the working flow of an artificial neuron as follows:

$$z = w_1 \cdot x_1 + \cdots + w_M \cdot x_M = \sum_{m=1}^{M} w_i \cdot x_i = \mathbf{w}^{\mathrm{T}} \cdot \mathbf{x}, \tag{2.1}$$

which represents the integration of the input signals arriving at the cell. The right-hand side of (2.1) represents the conventional scalar product between vectors. Subsequently, the value of $z$ is processed at the cell nucleus by the decision function $\phi(z)$ as follows:

$$\phi(z) = \begin{cases} 1, & \text{if } z \geq \theta, \\ -1 & \text{if } z < \theta, \end{cases} \tag{2.2}$$

where $\theta$ is referred to as the threshold value. Conventionally, this value is put on the left-hand side of (2.1), yielding the following:

$$z = w_0 \cdot x_0 + w_1 \cdot x_1 + \cdots + w_M \cdot x_M = \sum_{m=0}^{M} w_i \cdot x_i = \mathbf{w}^{\mathrm{T}} \cdot \mathbf{x}, \tag{2.3}$$

and

$$\phi(z) = \begin{cases} 1, & \text{if } z \geq 0, \\ -1 & \text{if } z < 0, \end{cases} \tag{2.4}$$

where $x_0 = 1$ and $w_0 = -\theta$, which is usually referred as bias unit.

Based on this representation, Frank Rosenblatt introduced the first concept of the perceptron learning rule. In particular, he proposed an algorithm to learn the optimal value for the weight coefficient in the context of supervised learning regression and classification.

## 2.2 Neural network architectures

Now that we have a clear understanding of what an artificial neuron is and how it is mathematically modelled, we can introduce the concept of neural networks and deep neural networks (DNNs). Although these terms are used interchangeably, some differences persist between them. To provide a high-level definition, a neural network mimics the way the biological neurons in the human brain work, whereas a DNN comprises several layers of neural networks. In other words, a neural network comprises an input layer which get the raw inputs, one hidden layer consisting of several artificial neurons fully connected with the input layer, and an output layer also represented by artificial neurons which is fully connected with the hidden layer. Once such structure contains more than one hidden layer, we start to talk about DNN.

A typical DNN architecture includes an input layer, an output layer, and multiple hidden layers. The input and output layers receive information and produce the final network predictions or decisions. The hidden or invisible layers extract patterns and increasingly complex features from the input data. Due to their inner structure, DNNs encompass various architectures designed for specific tasks and can be broadly categorised into five main classes:

- **Feedforward neural networks (FNN)**: Commonly used for tasks like classification and regression, these represent the most basic type of DNNs, where information flows in one direction, i.e. from the input layer through the hidden layers to the output layer.
- **Convolutional neural networks (CNN)**: These have been highly successful in image recognition, computer vision, and related tasks. Indeed, CNNs are designed for processing structured grid data, such as images, by utilising convolutional layers to automatically learn hierarchical representations of features.
- **Recurrent neural networks (RNN)**: RNNs are designed to handle sequential data and have connections that form cycles, allowing them to capture temporal dependencies. They are commonly used in natural language processing, speech recognition, and time-series analysis.
- **Long short-term memory networks (LSTM) and gated recurrent units (GRU)**: These represent specialised RNNs designed and with proven efficiency on tasks which need to capture long-term dependencies in sequential data.
- **Generative adversarial networks (GAN)**: Composed of a generator and a discriminator network trained simultaneously through adversarial learning, GANs are used to generate new data instances that closely resemble a given dataset. Their applications include image synthesis, style transfer, and data generation.

This categorisation represents a broad overview about each type of DNN, highlighting how the choice of a DNN architecture depends on the nature of the task, the characteristics of the data, and the desired outcomes. More detailed discussions and explanations about each architecture can be found in the current literature.

## 2.3 Choosing activation functions for multi-layer neural networks

When illustrating the structure of an artificial neuron in Section 2.1, it has been shown how the neuron nucleus is modelled through the following binary activation function:

$$\phi(z) = \begin{cases} 1, & \text{if } z \geq 0, \\ -1 & \text{if } z < 0, \end{cases} \tag{2.5}$$

where $z = \mathbf{w}^T \cdot \mathbf{x}$. However, this represents only one case since technically it is possible to choose any type of activation function. As we will see later in the context of training a DNN, one of the first algorithms for neural network training is based on the assumption that a linear function is used as activation function. Although this can provide advantages in terms of training algorithms, in practice, it would not be useful to use linear activation functions for both hidden and output layers. This is mainly because we want to introduce non-linearity in order to be able to deal with complex

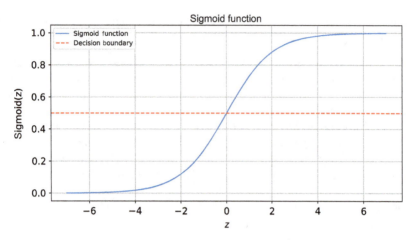

Figure 2.2  Sigmoid function

real-life problems. Due to their fundamental role in DNN design, some commonly used activation functions will be briefly discussed below.

### 2.3.1 Sigmoid function

The sigmoid activation function, also known as logistic function, maps any real-valued number to a range between 0 and 1 through a smooth S-shaped curve (see Figure 2.2). Mathematically, it is defined as follows:

$$\phi(z) = \frac{1}{1+e^{-z}}. \tag{2.6}$$

This type of function is commonly employed as the activation function of the output layer of binary classification models. Indeed, since constrained between 0 and 1, output values can be interpreted as probabilities that a particular sample belongs to a class or to another. If the output is close to 0, it signifies a low probability, while an output close to 1 indicates a high probability. For this reason, the 0.5 value is often referred as boundary value. However, even it has been historically popular, the sigmoid function has some limitations, such as susceptibility to vanishing gradient problems during training, especially.

### 2.3.2 Hyperbolic tangent function

The hyperbolic tangent function, commonly referred to as tanh, is another important activation function in the domain of DNNs.

$$\phi(z) = \frac{e^z - e^{-z}}{e^z + e^{-z}}. \tag{2.7}$$

From its definition in (2.7), one can notice how it exhibits non-linear behaviour, essential to model intricate patterns and relationships within the data. As illustrated in Figure 2.3, the input values are squashed to a bounded output range, i.e. $(-1; 1)$,

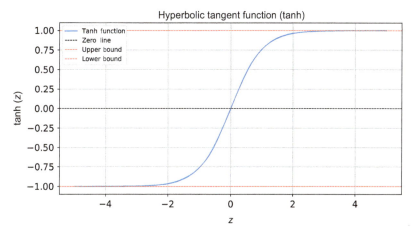

*Figure 2.3  Hyperbolic tangent function*

which makes it particularly suitable for scenarios where zero-centred outputs are desired. This zero-centred property can be advantageous in certain situations, aiding in the mitigation of the vanishing gradient problem, a challenge often encountered during the training of deep networks. The tanh function finds application in hidden layers of neural networks, where its ability to model complex relationships and its zero-centred output contribute to the network's learning capabilities. It is commonly used in RNNs and LSTMs for tasks involving sequential data, and it is often a suitable alternative to the sigmoid function, even if it still presents some susceptibility to vanishing gradient problems. In addition, its wider output range can enhance the convergence speed during the training of DNNs.

### 2.3.3  Rectified linear unit

The rectified linear unit (ReLU), defined as $ReLU(z) = \max(0, z)$, introduces non-linearity, as well as simplicity and computational efficiency. It allows positive values to pass through unchanged, while setting negative values to zero (see Figure 2.4). Compared to sigmoid and tanh, ReLU often exhibits faster convergence during training due to its non-saturating nature. This allows the ReLU to mitigate the issue of vanishing gradient by providing strong gradients for positive inputs, promoting in turn more efficient weight updates and then faster learning. ReLU's main usage is in hidden layers of DNNs, particularly in CNNs and deep feedforward networks. Its simplicity and ability to accelerate training make it a popular choice for various tasks such as image classification, object detection, and natural language processing. However, despite its proven effectiveness in many DNN architectures, ReLU has certain issues. Two notable problems associated with ReLU are the 'dying ReLU' problem and the potential for exploding gradients in certain scenarios. In the first case, this

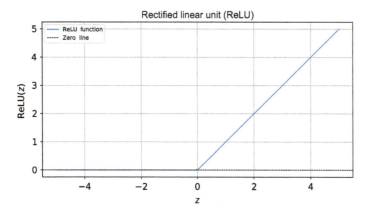

*Figure 2.4   Rectified linear unit (ReLU) function*

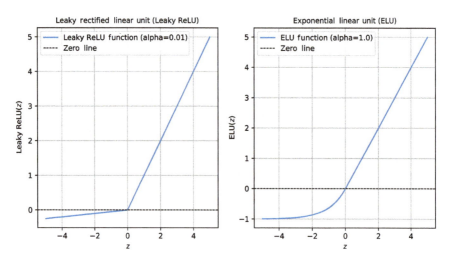

*Figure 2.5   Leaky ReLU and exponential linear unit (ELU) functions*

happens when, during training, the neurons using ReLU may become inactive (outputting zero) for all inputs. On the other hand, in very deep networks, the gradients during backpropagation can become extremely large (exploding gradients), leading to numerical instability and difficulties in convergence during training. Last but not least, ReLU is not differentiable at 0, which can cause optimisation challenges in gradient-based training algorithms. For this reason, the leaky ReLU and exponential linear unit (ELU) are alternative activation functions that aim to address these issues. In particular, based on their definitions and graphical representation in (2.8) and (2.9) and illustrated in Figure 2.5, one can notice how they address the dying ReLU problem by a small slope and smoother slope, respectively, for negative inputs with a small positive value $\alpha$. In addition, these new types of transition results are also helpful in reducing the exploding gradients problem, especially through the use of ELU as the activation function.

$$\phi(z) = \max(\alpha z, z) \tag{2.8}$$

$$\phi(z) = \begin{cases} z & \text{if } z > 0, \\ \alpha(e^z - 1) & \text{if } z \leq 0. \end{cases} \tag{2.9}$$

### 2.3.4 Softmax function

The softmax function is a fundamental activation function in the context of multi-class classification tasks in DNNs, where each neuron of the output layer provides the probability of a particular input belonging to a particular class. However, in contrast to using sigmoid function for each neuron, in this case the use of a softmax function guarantees that all the probabilities sum up to 1. Indeed, the probability that a particular input $z$ belongs to the $j$-th class is defined as follows:

$$\phi(z)_j = \frac{e^{z_j}}{\sum_{i=1}^{M} e^{z_i}}, \tag{2.10}$$

where $M$ represents the total number of classes, while $z_j = \mathbf{w}_j^T \cdot \mathbf{x}$, in which $\mathbf{w}_j$ represents the weights for combining the inputs at the neuron representing the $j$-th class. However, the softmax function is widely used and effective in many scenarios. Still, it does have certain limitations and considerations, such as sensitivity to outliers and high computational cost due to the exponentials and the normalisation.

## 2.4 Learning algorithm

A learning algorithm in DNN refers to the method used to adjust the weights of the connections between the neurons in the network. This section will discuss the most popular training algorithms.

### 2.4.1 Perceptron learning rule

As already mentioned in Section 2.1, based on the MCP threshold model, Frank Rosenblatt published the first concept of the perceptron learning rule. It can be considered as one of the first training algorithms for neural network. In particular, considering a single artificial neuron used for a binary classification task, we indicate with $\mathbf{X}$ the set containing the $M$ training samples. To the generic training sample $\mathbf{x}^{(i)} \in \mathbf{X}$ is associated a label $y^{(i)}$ that corresponds to an estimated value $\hat{y}^{(i)}$, and then a corresponding prediction error $\Delta^{(i)} = y^{(i)} - \hat{y}^{(i)}$. Now we can summarise the perceptron training algorithm as illustrated in Algorithm 2.1.

The update of the weights, indicating with $w_j$ the $j$-th component of the weight vector, is as follows:

$$w_j \longleftarrow w_j + \Delta w_j, \quad \Delta w_j = \eta \cdot \Delta^{(i)} \cdot x_j^{(i)}, \tag{2.11}$$

**Algorithm 2.1** Perceptron learning rule

**Require:** Initialise weights **w** with small random values
1: **for** $\mathbf{x}^{(i)} \in \mathbf{X}$ **do**
2:     Compute the output value $\hat{y}^{(i)}$
3:     Compute the estimation error $\Delta^{(i)} = y^{(i)} - \hat{y}^{(i)}$
4:     Update the weights
5: **end for**

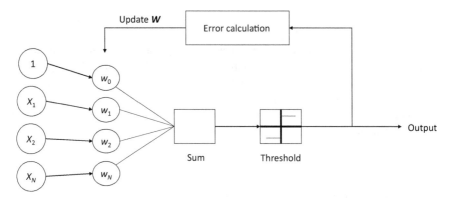

*Figure 2.6    Diagram of MCP neuron model*

where $\eta$ is the learning rate and is typically a constant value between 0 and 1. Then, it is important to note how all the weights are updated simultaneously for each training sample before a new estimation is performed. If the training samples has $N$ features, the updating procedure for the $i$-th training sample is performed as follows:

$$\Delta w_0 = w_0 + \eta \cdot (y^{(i)} - \hat{y}^{(i)})$$
$$\Delta w_1 = w_1 + \eta \cdot (y^{(i)} - \hat{y}^{(i)}) \cdot x_1^{(i)}$$
$$\vdots$$
$$\Delta w_N = w_N + \eta \cdot (y^{(i)} - \hat{y}^{(i)}) \cdot x_N^{(i)}$$

The entire learning rule can be summarised through Figure 2.6. Although the perceptron learning rule is quite simple, it has some drawbacks. In particular, it works well when the classes are linearly separable, i.e. the decision boundary can be expressed through a linear equation. Then, when possible, it is suggested to have a visual look at the data. If the linear separation condition is not satisfied, it is possible to set a maximum number of passes through the dataset, i.e. epochs, after which the algorithm will stop.

### 2.4.2  Adaptive linear neurons and the gradient descent learning method

The adaptive linear (Adaline) neuron model represents another type of neuron model proposed by Bernard Widrow and Tedd Hoff after Rosenblatt's perceptron model, which can be considered an improvement of the perceptron. The key difference between Adaline and the perceptron learning rule is that the former updates its weights based on a linear function. Indeed, the Adaline activation function $\phi(z)$ is the identity function:

$$\phi(z) = \mathbf{w}^T \cdot \mathbf{x}. \tag{2.12}$$

while the decision rule is always based on the step function. The main differences are summarised in Figure 2.7. Now, let us take a look at the main advantages that adopting Adaline brings to the context of learning algorithms. First of all, we should define an objective function that basically is a cost function that the designer wants to minimise during the learning process. In the case of Adaline, the cost function $J$ to learn the weights is the sum of squared error (SSE) between the calculated label and the true label class:

$$J(\mathbf{w}) = \frac{1}{2} \sum_i \left( y^i - \phi(z^i) \right)^2. \tag{2.13}$$

As one can note, the first main advantage of using this type of function is that it is differentiable. But more importantly, it is convex and allows the application of a very simple and powerful optimisation algorithm called **gradient descent** to find the optimal weights that minimise the cost function. In the context of gradient descent,

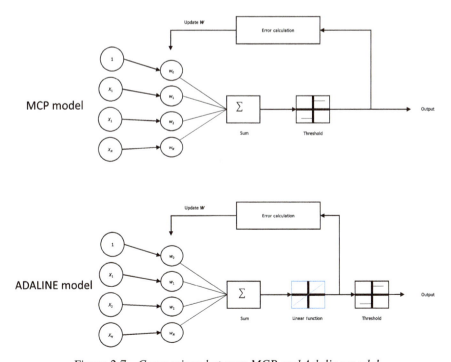

*Figure 2.7   Comparison between MCP and Adaline model*

the weights are updated by taking a step in the opposite direction of the gradient $\nabla J(\mathbf{w})$ of the cost function, as follows:

$$\mathbf{w} \leftarrow \mathbf{w} + \Delta \mathbf{w}, \tag{2.14}$$

where $\Delta \mathbf{w} = -\eta \cdot \nabla J(\mathbf{w})$. By computing the gradient of the cost function, i.e. the partial derivative along each component of $\mathbf{w}$, it is easy to show that:

$$\frac{\partial J}{\partial w_j} = -\sum_i \left( y^i - \phi(z^i) \right) \cdot x_j^{(i)}, \tag{2.15}$$

Then, for the generic $j$-th weight, we have the following update rule:

$$\Delta w_j = \eta \cdot \sum_i \left( y^i - \phi(z^i) \right) \cdot x_j^{(i)}. \tag{2.16}$$

### 2.4.3 Batch gradient descent

From what is illustrated so far, one can easily note how, compared to the perceptron learning rule, the weights are updated simultaneously in the Adaline. In addition, the update of the weights is performed on the whole training set. This type of learning procedure is referred to as **batch gradient descent**. Indeed, it is an optimisation algorithm that calculates the gradient of the loss function over the entire training set in several epochs. This means that for each epoch, the weights connecting the neuron with the input are updated in the way of the negative gradient to minimise the loss function.

### 2.4.4 Dealing with large training datasets: stochastic gradient descent

With the batch gradient descent, the main drawback comes when dealing with large datasets containing millions of data points, which is quite frequent in ML-related applications. Applying the batch gradient descent could result in quite an expensive and time-consuming procedure. In that case, a common alternative procedure is the stochastic gradient descent (SGD) procedure, also known as the online training procedure. In SGD, the weights are updated for each training sample by following only the part of the gradient related to that specific training sample, as illustrated below:

$$\Delta w_j = \eta \cdot \left( y^{(i)} - \hat{y}^{(i)} \right) \cdot x_j^{(i)}. \tag{2.17}$$

Due to its inner nature, the SGD is often considered as an approximation of the conventional gradient descent. The SGD comes with the main benefit of faster implementation thanks to the more frequent weight updates. In addition, it is more efficient in avoiding local minima of the cost function. Lastly, its online nature allows the neural network to train with new, fresh data as these arrives. However, it is important to guarantee good randomness in the data at each epoch.

## 2.4.5 A trade-off between gradient descent and SGD: Mini-batch gradient descent

A compromise between the batch gradient and the SGD is represented by the so-called **mini-batch gradient descent** algorithm. More specifically, this type of algorithm applies batch gradient descent but in a smaller subset of the training sample. In other words, the original dataset is split into small training subsets, and the weights are updated on the SSE of each subset.

## 2.4.6 SGD with momentum: overcoming the noisy nature of SGD and mini-batch

A major disadvantage of both SGD and mini-batch SGD algorithms is that the update of the weights can be very noisy. This, in turn, results in slow convergence due to the fact that these algorithms present many oscillations until they reach the minimum. To overcome these potential issues, the SGD with momentum introduces a momentum term, which is a moving average of gradients, to the standard SGD update rule. The main idea is to reduce the noise of the partial gradient by using exponential weighting average, meaning give more importance to recent updates compared to the previous update. This accelerates the convergence toward the relevant direction and reduces the fluctuation to the irrelevant direction. To understand better how this method works let us slightly change the notation to introduce a temporal dependence. Referring to the SGD update rule, we can rewrite the new update rule with temporal dependence as follows:

$$w_j(t) = w_j(t-1) - \eta \cdot \nabla J(\mathbf{w}(t)), \tag{2.18}$$

where $J(\mathbf{w}(t)) = \frac{1}{2}\left(y^{(i)} - \hat{y}^{(i)}(t)\right)^2$. The main idea behind momentum is to introduce a moving average of past gradients to the parameter updates, helping the update algorithm to maintain a more consistent direction toward the minimum. Denoting with $m_t$ the momentum value at time $t$ and with $\beta$ the parameter (usually close to 1, e.g. 0.9 or 0.99), the new update rule becomes

$$\begin{aligned} m_t &= \beta \cdot m_{t-1} + (1-\beta) \cdot \nabla J(\mathbf{w}(t)), \\ w_j(t+1) &= w_j(t) - \eta \cdot m_t. \end{aligned} \tag{2.19}$$

As one can notice from (2.19), the introduction of the momentum term allows the algorithm to ignore small fluctuations and continue to move in the overall direction of the gradient. More specifically, if the gradient points consistently in one direction, the momentum term will accumulate and reinforce the movement. This definitively improves the performances in terms of convergence time. However, this comes at the cost of computing one more variable for each update.

A variation of this algorithm is the **Nesterov accelerated gradient (NAG)** algorithm. This was introduced to obtain a momentum-based method more responsive to the curvature of the optimisation landscape, which can lead to faster convergence, especially in scenarios with high curvature. The main idea of NAG is to make a look-ahead before updating the parameters by calculating the gradient at a point slightly ahead in the direction of the momentum as follows:

$$m_t = \beta \cdot m_{t-1} + \eta \cdot \nabla J(\mathbf{w}(t) - \beta \cdot m_{t-1}),$$
$$w_j(t+1) = w_j(t) - m_t. \qquad (2.20)$$

### 2.4.7 Adjusting the learning rate: adaptive gradient descent (AdaGrad)

For all the SGD-based algorithms discussed so far, the learning rate remains constant and the same for all the parameters during the learning process. Under this perspective, the AdaGrad algorithm has been designed to adapt the learning rates of individual parameters during the training process, adjusting them based on the historical gradients of each parameter. More specifically, it allows larger updates for infrequently updated parameters and smaller updates for frequently updated ones. Before illustrating how this algorithm works, we must modify the notations adopted so far. We firstly need to define a matrix $G \in \mathbb{R}^{N \times N}$ which is a diagonal matrix in which each element contains the aggregated sum of the squares of the gradients with respect of each parameter more in detail:

$$G_{i,i}(t) = \sum_{\tau=1}^{t} \left( \frac{\partial J(\mathbf{w}(t))}{\partial w_i} \right)^2. \qquad (2.21)$$

Then, considering a small constant $\varepsilon$, the AdaGrad rule for updating the parameters in vector notation can be expressed as follows:

$$\mathbf{w}(t+1) = \mathbf{w}(t+1) - \eta \cdot \frac{1}{\sqrt{\varepsilon + G}} \odot \nabla(J(\mathbf{w}(t))), \qquad (2.22)$$

where the $\odot$ operator represents the point-wise multiplication of the elements. As can be seen from (2.21), all the squared terms are positive, meaning that the accumulated sum keeps on growing during training. Therefore, the learning rate keeps shrinking as the training continues, and it eventually becomes infinitely small, potentially leading to the vanishing gradient problem. Other algorithms like **Adadelta**, **RMSprop**, and **Adam** try to resolve this flaw.

## 2.5 Adjusting the deep neural network's coefficients through the error backpropagation

As explained at the beginning of the chapter, artificial neural networks have gained a lot of attention within the last two decades. However, after the MCP model presentation and then Rosenblatt's perceptron model implementation, there was a loss of interest from machine learning practitioners since no one had a clear good solution on how to train a neural network with multiple layers. This area remained undiscovered until 1986 when the backpropagation algorithm was proposed. This section briefly introduce some useful notations and definitions that can be used to illustrate this algorithm.

### 2.5.1 Multi-layer neural network

As illustrated in Figure 2.8, a multi-layer neural network is represented by an input layer, an output layer, and several hidden layers, each of them consisting of several artificial neurons. In the case there is more than one hidden layer, this is referred to as DNN. We will indicate with $a_i^l$ the $i$-th neuron (or activation element) of the hidden layer while $a_i^{in}$ and $a_i^{out}$ the $i$-th element of the input and output layer, respectively. In addition, we will suppose that all the layers have two elements. For the sake of easy explanation and without loss of generality, the following assumptions are made:

- $H = 1$ number of hidden layers;
- All the layers, i.e. input, hidden, and output, have two neurons.
- The neural network is fully connected;
- Only one pair input–output in the training is considered;
- $e_1$, $e_2$, and $E = e_1 + e_2$ represent the estimation and aggregated errors, respectively. Typically, $e_1$ and $e_2$ are calculated as a mean square error.

$w_{k,j}^l$ represents the weight coefficient between the $k$-th unit of the $(l+1)$-th layer with the $j$-th unit from layer $l$. These will all be part of a matrix $\mathbf{W}^l$. The main aim of the backpropagation algorithm is to calculate how the error varies with respect to each weight of the network in a recursive way, propagating from the end of the network to the beginning. To this end, this algorithm exploits the chain rule for calculating gradients of nested functions. In the case of a one-dimensional function, this rule is expressed as follows:

$$\frac{\partial f(g(z(x)))}{\partial x} = \frac{\partial f}{\partial g} \cdot \frac{\partial g}{\partial z} \cdot \frac{\partial z}{\partial x}. \tag{2.23}$$

Recall that each activation layer element can be expressed as follows:

$$a^l = \phi^l(z(w^l)), \tag{2.24}$$

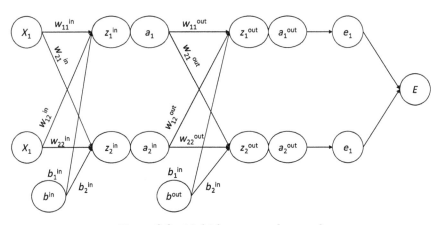

*Figure 2.8  Multi-layer neural network*

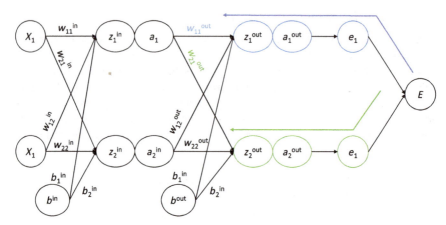

Figure 2.9 Backpropagation paths for calculating $\frac{\partial E}{\partial w_{11}^{out}}$ (blue) and $\frac{\partial E}{\partial w_{21}^{out}}$ (green)

with $\phi()$ being a generic activation function, we can proceed to extract the various partial derivatives from the output layer as illustrated in Figure 2.9:

$$\begin{aligned}\frac{\partial E}{\partial w_{11}^{out}} &= \frac{\partial e_1}{\partial a_1^{out}} \cdot \frac{\partial a_1^{out}}{\partial z_1^{out}} \cdot \frac{\partial z_1^{out}}{\partial w_{1,1}^{out}}, \\ \frac{\partial E}{\partial w_{12}^{out}} &= \frac{\partial e_1}{\partial a_1^{out}} \cdot \frac{\partial a_1^{out}}{\partial z_1^{out}} \cdot \frac{\partial z_1^{out}}{\partial w_{1,2}^{out}}, \\ \frac{\partial E}{\partial w_{21}^{out}} &= \frac{\partial e_2}{\partial a_2^{out}} \cdot \frac{\partial a_2^{out}}{\partial z_2^{out}} \cdot \frac{\partial z_2^{out}}{\partial w_{2,1}^{out}}, \\ \frac{\partial E}{\partial w_{22}^{out}} &= \frac{\partial e_2}{\partial a_2^{out}} \cdot \frac{\partial a_2^{out}}{\partial z_2^{out}} \cdot \frac{\partial z_2^{out}}{\partial w_{2,2}^{out}}.\end{aligned} \quad (2.25)$$

For better visualisation, the set of equations provided in (2.25) have been arranged in vectors as follows:

$$\begin{bmatrix} \frac{\partial E}{\partial w_{11}^{out}} & \frac{\partial E}{\partial w_{12}^{out}} \\ \frac{\partial E}{\partial w_{21}^{out}} & \frac{\partial E}{\partial w_{22}^{out}} \end{bmatrix} = \begin{bmatrix} \frac{\partial e_1}{\partial a_1^{out}} \\ \frac{\partial e_1}{\partial a_2^{out}} \end{bmatrix} \odot \begin{bmatrix} \frac{\partial a_1^{out}}{\partial z_1^{out}} \\ \frac{\partial a_2^{out}}{\partial z_2^{out}} \end{bmatrix} \odot \begin{bmatrix} \frac{\partial z_1^{out}}{\partial w_{1,1}^{out}} & \frac{\partial z_1^{out}}{\partial w_{1,2}^{out}} \\ \frac{\partial z_2^{out}}{\partial w_{2,1}^{out}} & \frac{\partial z_2^{out}}{\partial w_{2,2}^{out}} \end{bmatrix}, \quad (2.26)$$

in which $\odot$ represents the point-wise multiplication. To provide a more compact form, we use the following notation:

$$\frac{\partial E}{\partial \mathbf{W}^{out}} = \delta^L \odot \frac{\partial \mathbf{Z}^{out}}{\partial \mathbf{W}^{out}}. \quad (2.27)$$

As we will see shortly, $\delta^L$ is something we can calculate and keep as it will appear later in the backward equations. Continuing in the same way, we can derive the gradient of the cost function with respect to the first set of coefficients. In this case,

*Deep neural networks* 29

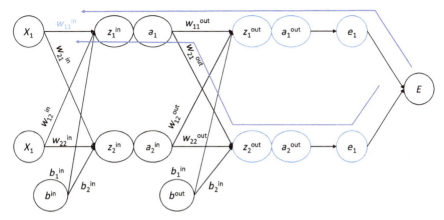

*Figure 2.10 Backpropagation paths for calculating $\frac{\partial E}{\partial w_{11}^{in}}$. This is obtained as the superposition of the contribution from two distinct paths*

as illustrated in Figure 2.10, we need to notice that these coefficients have influence in different network paths. Based on this, we can write the next set of equations:

$$\frac{\partial E}{\partial w_{1,1}^{in}} = \left( \frac{\partial e_1}{\partial a_1^{out}} \cdot \frac{\partial a_1^{out}}{\partial z_1^{out}} \cdot \frac{\partial z_1^{out}}{\partial a_1^{in}} + \frac{\partial e_2}{\partial a_2^{out}} \cdot \frac{\partial a_2^{out}}{\partial z_2^{out}} \cdot \frac{\partial z_2^{out}}{\partial a_1^{in}} \right) \cdot \frac{\partial a_1^{in}}{\partial z_1^{in}} \cdot \frac{\partial z_1^{in}}{\partial w_{1,1}^{in}}$$

$$\frac{\partial E}{\partial w_{1,2}^{in}} = \left( \frac{\partial e_1}{\partial a_1^{out}} \cdot \frac{\partial a_1^{out}}{\partial z_1^{out}} \cdot \frac{\partial z_1^{out}}{\partial a_1^{in}} + \frac{\partial e_2}{\partial a_2^{out}} \cdot \frac{\partial a_2^{out}}{\partial z_2^{out}} \cdot \frac{\partial z_2^{out}}{\partial a_1^{in}} \right) \cdot \frac{\partial a_1^{in}}{\partial z_1^{in}} \cdot \frac{\partial z_1^{in}}{\partial w_{1,2}^{in}}$$

$$\frac{\partial E}{\partial w_{2,1}^{in}} = \left( \frac{\partial e_1}{\partial a_1^{out}} \cdot \frac{\partial a_1^{out}}{\partial z_1^{out}} \cdot \frac{\partial z_1^{out}}{\partial a_2^{in}} + \frac{\partial e_2}{\partial a_2^{out}} \cdot \frac{\partial a_2^{out}}{\partial z_2^{out}} \cdot \frac{\partial z_2^{out}}{\partial a_2^{in}} \right) \cdot \frac{\partial a_2^{in}}{\partial z_2^{in}} \cdot \frac{\partial z_2^{in}}{\partial w_{2,1}^{in}}$$

$$\frac{\partial E}{\partial w_{2,2}^{in}} = \left( \frac{\partial e_1}{\partial a_1^{out}} \cdot \frac{\partial a_1^{out}}{\partial z_1^{out}} \cdot \frac{\partial z_1^{out}}{\partial a_2^{in}} + \frac{\partial e_2}{\partial a_2^{out}} \cdot \frac{\partial a_2^{out}}{\partial z_2^{out}} \cdot \frac{\partial z_2^{out}}{\partial a_2^{in}} \right) \cdot \frac{\partial a_2^{in}}{\partial z_2^{in}} \cdot \frac{\partial z_2^{in}}{\partial w_{2,2}^{in}}$$

(2.28)

From this new set of equations, it is easy to notice the presence of the values from $\delta$ previously calculated. This will allow to rewrite (2.28) in a more compact form:

$$\frac{\partial E}{\partial w_{1,1}^{in}} = \left( \delta_1^L \cdot \frac{\partial z_1^{out}}{\partial a_1^{in}} + \delta_2^L \cdot \frac{\partial z_2^{out}}{\partial a_1^{in}} \right) \cdot \frac{\partial a_1^{in}}{\partial z_1^{in}} \cdot \frac{\partial z_1^{in}}{\partial w_{1,1}^{in}}$$

$$\frac{\partial E}{\partial w_{1,2}^{in}} = \left( \delta_1^L \cdot \frac{\partial z_1^{out}}{\partial a_1^{in}} + \delta_2^L \cdot \frac{\partial z_2^{out}}{\partial a_1^{in}} \right) \cdot \frac{\partial a_1^{in}}{\partial z_1^{in}} \cdot \frac{\partial z_1^{in}}{\partial w_{1,2}^{in}}$$

$$\frac{\partial E}{\partial w_{2,1}^{in}} = \left( \delta_1^L \cdot \frac{\partial z_1^{out}}{\partial a_2^{in}} + \delta_2^L \cdot \frac{\partial z_2^{out}}{\partial a_2^{in}} \right) \cdot \frac{\partial a_2^{in}}{\partial z_2^{in}} \cdot \frac{\partial z_2^{in}}{\partial w_{2,1}^{in}}$$

$$\frac{\partial E}{\partial w_{2,2}^{in}} = \left( \delta_1^L \cdot \frac{\partial z_1^{out}}{\partial a_2^{in}} + \delta_2^L \cdot \frac{\partial z_2^{out}}{\partial a_2^{in}} \right) \cdot \frac{\partial a_2^{in}}{\partial z_2^{in}} \cdot \frac{\partial z_2^{in}}{\partial w_{2,2}^{in}}$$

(2.29)

With a further re-arrangement, we can write everything in the following form:

$$\begin{bmatrix} \frac{\partial E}{\partial w_{11}^{in}} & \frac{\partial E}{\partial w_{12}^{in}} \\ \frac{\partial E}{\partial w_{21}^{in}} & \frac{\partial E}{\partial w_{22}^{in}} \end{bmatrix} = \begin{bmatrix} \delta_1^L \\ \delta_2^L \end{bmatrix}^T \cdot \begin{bmatrix} \frac{\partial z_1^{out}}{\partial a_1^{in}} & \frac{\partial z_1^{out}}{\partial a_2^{in}} \\ \frac{\partial z_2^{out}}{\partial a_1^{in}} & \frac{\partial z_2^{out}}{\partial a_2^{in}} \end{bmatrix} \odot \begin{bmatrix} \frac{\partial a_1^{in}}{\partial z_1^{in}} \\ \frac{\partial a_2^{in}}{\partial z_2^{in}} \end{bmatrix} \odot \begin{bmatrix} \frac{\partial z_1^{in}}{\partial w_{1,1}^{in}} & \frac{\partial z_1^{in}}{\partial w_{1,2}^{in}} \\ \frac{\partial z_2^{in}}{\partial w_{2,1}^{in}} & \frac{\partial z_2^{in}}{\partial w_{2,2}^{in}} \end{bmatrix}. \quad (2.30)$$

Using again the compact form we can write (2.30) as follows:

$$\frac{\partial E}{\partial \mathbf{W}^{in}} = (\delta^L)^T \cdot \frac{\partial Z^{out}}{\partial A^{in}} \odot \frac{\partial A^{in}}{\partial z^{in}} \odot \frac{\partial Z^{in}}{\partial \mathbf{W}^{in}}. \quad (2.31)$$

Finally, having a close look at the network structure, it is easy to note the value of some derivatives, which can allow us to write the final version of (2.31) as follows:

$$\frac{\partial E}{\partial \mathbf{W}^{in}} = (\delta^L)^T \cdot \mathbf{W}^{out} \odot \frac{\partial A^{in}}{\partial z^{in}} \odot \mathbf{X}^{in}, \quad (2.32)$$

which is easy to generalise by induction for a generic layer $h$ when $H > 1$:

$$\frac{\partial E}{\partial \mathbf{W}^h} = (\delta^{(h+1)})^T \cdot \mathbf{W}^{(}h+1) \odot \frac{\partial A^h}{\partial z^h} \odot \mathbf{X}^h. \quad (2.33)$$

The reader can also easily obtain the equations necessary to adjust the bias factors. Due to this powerful algorithm, it is possible to calculate all the error gradients with respect to the network weights and adjust them using one of the SGD methods explained in the previous section of this chapter. To date, many dedicated and free-to-use software libraries for these types of operations are provided. One of the most popular is TensorFlow [1].

## 2.6 Overfitting and underfitting

At this point, we have described what DNNs are, their main structures, building blocks, and how to adjust the relative weights during the training. However, another important aspect that needs to be taken into account during the training session, especially in the context of supervised learning, is the capability of the considered DNN-based model to *generalise* from the training set, i.e. its ability to make accurate predictions on unseen data. Usually, a best practice is to divide the entire dataset into training and test sets, where the latter is used to check the generalisation aspects of the designed model.

Typically, the considered model is designed to provide precise predictions on the training set. If the training set and the new unforeseen datasets share sufficient commonalities, there will be a high chance that the model will exhibit accuracy on new data as well. However, in many cases, we can observe deviations from our expectations when using unforeseen data. This happens when the model we are considering is too complex, meaning that it takes all the details (features) of the training set into full consideration. This is referred to as **overfitting**. In other words, overfitting takes place when a model is tailored too closely to the intricacies of the training set, resulting in a model that performs effectively on the training data but struggles to generalise to new unseen data. On the other hand, if it is too simple, you might not be able to capture all the essential aspects and variability in the data, which leads to a model badly performing on the training set. Choosing too simple a model is called **underfitting**.

In literature, these terms are also referred to as bias-variance trade-off. To understand this concept, we need to consider the mathematical expression of the prediction error. Let us suppose that we want to predict a variable $Y$ that can be modelled as $Y = f(x) + e$, where the last term represents the error between the model and the *true* function representing the data, which typically follows a normal distribution with 0 mean. Based on these assumptions, we will build a model $\hat{f}(x)$ modelling $f(x)$. At this point, we can write the mean squared error between our model and the *ideal* data model as

$$Err(x) = E\left[(Y - \hat{f}(x))^2\right], \tag{2.34}$$

which, after some simple mathematical transformations, can be expressed as

$$Err(x) = \left(f(x) - E[\hat{f}(x)]\right)^2 + E\left[\left(f(x) - \hat{f}(x)\right)^2\right] + \sigma_e. \tag{2.35}$$

The first term on the right side of (2.35) is referred to as **bias**, i.e. the difference between the average prediction of our model and the correct value which we are trying to predict. The middle term represents the **variance**, which is the variability of model prediction for a given data point or a value. The last term is the variance of the error. Models with high bias pay very little attention to the training data and oversimplify the model, leading to high training and test data errors. On the other hand, models with high variance pay a lot of attention to training data and do not generalise the data which has not been seen before.

Then, the main ability of the model designer stays in finding a model $\hat{f}(x)$ with both low variance and low bias, i.e. a model able to reach the optimal trade-off between its inner complexity and accuracy as illustrated in Figure 2.11. In this

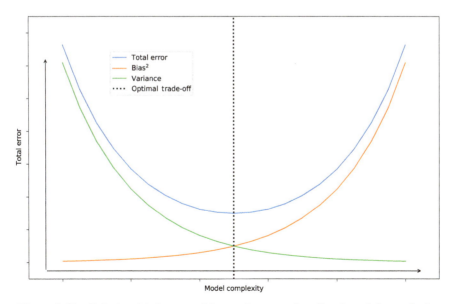

*Figure 2.11   Relationship between bias-variance trade-off and model complexity*

context, different techniques can be used to assess the generalisation of the model. These techniques, which range from the pre-processing of datasets to the introduction of regularisation terms within the model, which are outside the scope of this book. Readers interested in the topic are encouraged to read [2].

## 2.7 Summary

This chapter has provided a brief introduction to the concept of DNNs. In particular, it introduced the concept of neurons and the first MCP model, which started the development of this area. Subsequently, different architectures of DNN have been illustrated, as well as the different types of activation functions commonly used. After providing the basics on the main building blocks of a DNN, the chapter illustrated the optimisation algorithms that can adjust the network weights. Final remarks on model overfitting and model underfitting have also been provided.

## References

[1] Abadi M, Agarwal A, Barham P, *et al*. TensorFlow: Large-Scale Machine Learning on Heterogeneous Systems; 2015. Software available from tensorflow.org. Available from: https://www.tensorflow.org/.

[2] Shalev-Shwartz S and Ben-David S. *Understanding Machine Learning: From Theory to Algorithms*. Cambridge University Press; 2014.

*Part II*
# Deep reinforcement learning

*Chapter 3*
# Markov decision process

Deep reinforcement learning (DRL) merges RL with DL, enabling the resolution of intricate decision-making challenges that were once beyond the capabilities of machines. Consequently, this amalgamation unlocks many possibilities for applications in various domains, including healthcare, robotics, smart grids, finance, and more. In this chapter, we will provide an overview of DRL models, algorithms, and techniques by giving a description of RL as Markov decision processes.

## 3.1 A brief recap on reinforcement learning: a Markovian perspective

As illustrated in the previous chapters, reinforcement learning (RL) is a subfield of machine learning dedicated to sequential decision-making. The primary objective of RL is to design an agent that interacts with its environment to discover the optimal sequence of actions leading to the best possible outcome. The agent observes the state of the environment and, based on its policy, takes an action. The resulting feedback from the environment is then utilised to update the policy, aiming for optimality. In other words, the agent learns how to choose actions strategically to maximise the rewards received from the environment. To this end, it is worth mentioning that RL uses trial-and-error experience, meaning that, unlike other ML techniques, such as supervised learning, the agent does not require complete knowledge or control of the environment. Instead, it only needs to be able to interact with the environment and collect information.

For the sake of providing a more formal definition, RL can be modelled as a discrete-time stochastic process with the following working process:

- The agent starts by observing the environment with an initial state $s_0 \in \mathcal{S}$, through an initial observation $\omega_0 \in \Omega$;
- At each time step $t$, the agent takes an action $a_t \in \mathcal{A}$;
- Following the taken action the agent receives a reward $r_t \in \mathcal{R}$ from the environment;
- The action also causes the environment moving to state $s_{t+1} \in \mathcal{S}$, perceived by the agent through an observation $\omega_{t+1} \in \Omega$.

This represents the first working flow of a RL mechanism firstly introduced by Bellman and later extended by Barto [1].

For those familiar with control theory, it can be easily noted how the working flow of an RL-based algorithm is modelled as a Markov decision process (MDP). In particular, indicating with $\mathscr{P}(a|b)$ the conditional probability that the event $a$ is conditioned by the event $b$, a discrete-time stochastic RL process is said to be Markovian if it satisfies the following conditions:

$$\begin{aligned} \mathscr{P}(\omega_{t+1}|\omega_t, a_t) &= \mathscr{P}(\omega_{t+1}|\omega_t, a_t, \cdots, \omega_0, a_0); \\ \mathscr{P}(r_t|\omega_t, a_t) &= \mathscr{P}(r_t|\omega_t, a_t, \cdots, \omega_0, a_0). \end{aligned} \quad (3.1)$$

In other words, it means that for an MDP, the future of the process only depends on the current observation, i.e. the agent is not influenced and is not interested in looking at the full history. Based on this property, the MDP [2] is defined by a 5-tuple $<\mathscr{S}, \mathscr{A}, \mathscr{P}, \mathscr{R}, \gamma>$ where $\mathscr{S}$ and $\mathscr{A}$ are the finite sets of states and actions, respectively; $\mathscr{P}$ is the transition probability function with $\mathscr{P}_{ss'}(a)$ representing the Probability to transition from state $s \in \mathscr{S}$ to state $s' \in \mathscr{S}$ after the action $a \in \mathscr{A}$ is taken; $\mathscr{R}$ is the reward obtained after the action $a$ is executed, and $\gamma$ represents the discount factor. It is worth mentioning that, within a MDP, the underlying considered system (environment) is fully observable at each time $t$, meaning that the state of the environment coincides with the observation, i.e. $s_t = \omega_t$.

### 3.1.1 Policy classification and expected reward

As already mentioned in Chapter 1, a RL-based algorithm is aimed at finding an optimal policy for the agent, which defines how the agent select actions based on the observed state. In this context, policies can be classified into either stationary or non-stationary according to whether they depend on the considered time-step or not. A non-stationary policy is used for finite-horizon cases where the agent seeks to optimise the cumulative rewards on a finite number of future time steps. However, only stationary policies are considered within this book. In addition, policies can also be classified as deterministic or stochastic. In the first case, the agent will have a clear view of what action needs to be taken when observing a particular case. Indicating the policy as $\pi()$, a deterministic policy can be expressed as

$$\pi(s) : \mathscr{S} \to \mathscr{A}. \quad (3.2)$$

On the other hand, a stochastic policy provides the probabilities that an action $a$ may be chosen in state $s$, which has the following mathematical expression:

$$\pi(s, a) : \mathscr{S} \times \mathscr{A} \to [0, 1]. \quad (3.3)$$

Let $\Pi$ be the set of all possible policies and $\mathscr{R}(s, a, s')$ be the reward obtained by taking the action $a$ in a state $s$ and transitioning in a state $s'$. Within this book we will consider the case of a RL agent whose goal is to find the stochastic policy $\pi(s, a) \in \Pi$

which maximises the value function $V^\pi(s) : \mathscr{S} \to \mathbb{R}$, also known as the expected return, such that:

$$V^\pi(s) = \mathbb{E}\left[\sum_{k=0}^{\infty} \gamma^k r_{t+k+1} | s_t = s, \pi\right], \qquad (3.4)$$

where $r_t = \mathbb{E}\left[\mathscr{R}(s_t, a_t, s_{t+1})\right]_{a_t \sim \pi(s,a)}$, represents, at time $t$, the average reward of being in a state $s_t$, taking an action $a_t$ and moving towards a new state $s_{t+1}$. The optimal expected return to which the agent aims is expressed as:

$$V^*(s) = \max_{\pi \in \Pi} V^\pi(s). \qquad (3.5)$$

### 3.1.2 Bellman's equation

In the context of RL, in addition to the $V$-value function defined in (3.4), another important function is the $Q$-value function, which is a function $\mathscr{Q}^\pi(s,a) : \mathscr{S} \times \mathscr{A} \to \mathbb{R}$ defined as follows:

$$Q^\pi(s,a) = \mathbb{E}\left[\sum_{k=0}^{\infty} \gamma^k r_{t+k+1} | s_t = s, a_t = a, \pi\right], \qquad (3.6)$$

which represents the average expected rewards through a policy $\pi$ which takes an action $a$ when in a particular state $s$. Also in this case, the optimal $Q$-value function can be defined as follows:

$$Q^*(s,a) = \max_{\pi \in \Pi} Q^\pi(s,a), \qquad (3.7)$$

from which one can easily notice that the following relationship exists between the $V$-value and $Q$-value functions:

$$V^*(s) = \max_{a \in \mathscr{A}} Q^*(s,a). \qquad (3.8)$$

Next, we can extract the optimal policy $\pi^*$ by choosing the action $a$ that gives maximum reward $Q^*(s,a)$ for state $s$. In mathematical form:

$$\pi^*(s) = \arg\max_{a \in \mathscr{A}} Q^*(s,a). \qquad (3.9)$$

The $Q$-value has the property that it can be recursively expressed through Bellman's equation. In particular, Bellman's equation expresses the $Q$-value of the current state and the $Q$-value of future states as follows:

$$Q(s,a) = R(s,a) + \gamma \max_{a' \in \mathscr{A}} Q(s',a'). \qquad (3.10)$$

In other words, this equation tells us that the maximum future reward is the reward the agent received in the current state $s$ through the action $a$ plus the maximum future reward in the future state $s'$ with respect to $a'$.

In summary, we can say that in an MDP, the optimal $V$-value function $V^*(s)$ represents the expected discounted reward in a given state $s$ when the agent follows a policy $\pi$. In contrast, the optimal $Q$-value $Q^*(s,a)$ is the expected discounted return when the agent is in a state $s$ and performs an action $a$ following the policy $\pi$. Another useful function is the advantage function $A(s, a) = Q(s, a) - V(s)$, which represents the goodness of taking the action $a$ compared to the expected return in the state $s$ when following the policy $\pi$.

### 3.1.3 Agent vs. environment: learning the policy

From what has been mentioned so far, it is clear that a RL-based method can be described through a MDP and that the final goal is to find the optimal policy that maximises each possible state's long-term reward. In doing so, a RL agent will involve one or more of the following components:

- a representation of the V-value function;
- a direct representation of the policy;
- a model of the environment, i.e. the transition probability function and the estimated reward function, in conjunction with a planning algorithm.

We can classify the RL into **model-free** RL and model-based RL. In particular, we talk about model-free RL when only the V-value function and the policy representation are introduced. If also the model of the environment is included, we will talk about **model-based** RL. We will cover these in the next chapters. Once the type of learning is defined, we can identify two possible ways to learn an optimal sequential decision-making policy:

- **Offline** learning, usually referred to as batch setting, which is mainly performed when only a limited amount of data is available for the considered environment;
- **Online** learning, where the agent gradually gathers experience in the environment in which it operates.

In both cases, all the training algorithms discussed in the upcoming chapters can be applied. The only substantial difference is how the agent is trained.

In the first case, the main characteristic of using a batch setting consists of collecting data, which will subsequently be used as a batch for training the agent and deploying it. In this process, the agent does not interact with the environment during the learning phase; instead, it learns from the fixed dataset. In addition, adopting offline learning can be advantageous when collecting real-time data is expensive or time consuming. However, this comes with some drawbacks such as (i) limited adaptability of the agent to changes that occurred in the environment after the data was collected, (ii) limited ability to discover optimal policies in regions of the state–action space that were not well represented in the initial dataset.

By adopting an online training process instead, the agent acquires data sequentially by constantly interacting with the environment over time, obtaining new data points that can used for learning. In this way, the agent updates its policy or V-value function incrementally as new data becomes available, allowing the agent to adapt to changes in the environment and improve its performance over time, making online training well suited for dynamic environments where the data distribution may change over time. This training approach addresses the adaptability issues of the offline training approach, but it can lead to some safety concerns since the agent is learning in a real-world environment where sub-optimal actions can have consequences. In addition, because of the possibility for the agent to gather experience through an active interaction with the environment, the exploration/exploitation strategy needs to be carefully managed. Indeed, since the agent needs to explore and gather data from the environment, an efficient exploration/exploitation might involve the actuation of sub-optimal actions that could negatively impact the learning process, as well as largely increase the convergence time.

In summary, the choice between offline and online policy learning is subject to a trade-off between adaptability and data efficiency. Usually, hybrid approaches, such as incorporating a pre-training phase with offline data followed by online fine-tuning, are used to mitigate some of these drawbacks and strike a balance between the advantages of both approaches.

## 3.2 Multi-arm bandit

The design of RL-based mechanisms can be categorised as a decision-making scenario. Indeed, these scenarios consist of an agent facing a set of options, which must be strategically chosen to maximise its cumulative reward over time. This type of setup is usually referred to as a multi-arm bandit (MAB) problem. The origin of this term lies in its metaphorical association with a bandit, analogous to an agent situated in front of a line of slot machines. The agent, with multiple arms, is tasked with determining not only which machines to engage with but also the frequency of interactions with each machine, the sequence in which to play them, and the decision of whether to persist with the current machine or explore alternatives. In other words, each arm represents either a decision or action, and the agent's goal is to iteratively learn the true reward distribution associated with each arm while optimising its choices in real time to maximise its long-term reward. This possibility for the agent to gather experience through its interaction with the environment, especially in an online setting, poses the challenge of the exploration–exploitation dilemma described subsequently.

### 3.2.1 Exploration and exploitation

The exploration–exploitation dilemma represents one of the main characteristics during the training of RL-based algorithms. Indeed, due to its inner nature, and for an agent to start to accumulate knowledge about the environment in which it is placed, it needs to perform interactively two complementary actions:

- **Exploration**: It is about obtaining information from the environment, like how transitions occur from one state to another and the reward amount obtained when an action is performed in different states.
- **Exploitation**: Based on the knowledge obtained during the exploration phase, the exploitation aspect of an RL-based agent is to literally exploit this knowledge to maximise the expected reward in each state, i.e. performing the action that provides the maximum reward.

However, even if these two phases are complementary, at a certain point, the agent has to face the dilemma of striking a balance between gaining further insights about its environment through exploration and pursuing the seemingly most promising strategy based on the accumulated experience through exploitation. For example, the first phase of interaction with the environment can be used for exploration purposes, while the agent will only try to maximise its own reward in the remaining part. However, this part is more delicate than it looks, especially when the agent operates in a continuous state–action space. Indeed, in such context, the agent must verify that the less-explored regions of the environment do not hold significant promise in terms of reward (exploration). At the same time, it has to accumulate experience and enhance its understanding of the dynamics in the areas of the environment that seem to be more promising in terms of reward (exploitation). Although different exploration techniques have been investigated, in this chapter we only illustrate the most commonly used in the majority of applications. Basically, these represent how to select an action in each state.

However, before moving in illustrating these methods, for the sake of better understanding let us first provide the definition of greedy policy. In doing so, we suppose that during time step $t$, we can choose an action $a_t$ from a set $\mathscr{A}$. For such an action, we can define an estimation of its quality at the time step $t$ as $Q_t(a)$, which should represent the expected reward. Supposing that we can have access to this estimation, the greedy policy consists of choosing every time the action which has the maximum expected reward:

$$a_t = \arg\max_{a \in \mathscr{A}} Q(a). \tag{3.11}$$

Clearly, this type of approach offers a low level of exploration (exploring new strategies) and a very high level of exploitation (maximising the long-term reward).

## 3.2.2 Epsilon-greedy method

A simple alternative is to behave greedily most of the time, but every once in a while, say with small probability $\varepsilon$, instead select randomly from among all the actions with equal probability, independently of the action-value estimates. This is what the $\varepsilon$-greedy methods do. By introducing a parameter $\varepsilon$, this strategy governs the exploration probability, offering a clear decision rule that strikes a balance between trying out new strategies and exploiting the current best-known strategy:

$$a_t = \begin{cases} \text{random action,} & \text{with probability } \varepsilon, \\ \arg\max_{a \in \mathscr{A}} Q(a), & \text{with probability } 1 - \varepsilon. \end{cases} \tag{3.12}$$

It is necessary to meticulously fine-tune the value of $\varepsilon$ based on the application's requirements and to carefully weigh the cost of exploration against the potential benefits of exploitation. This method stands as a practical and widely applicable approach to handling the exploration–exploitation dilemma.

## 3.2.3 Softmax selection

While $\varepsilon$-greedy action selection proves effective and is widely used for balancing exploration and exploitation in RL, a notable drawback of this method arises when exploration occurs, as it uniformly selects among all actions. This implies an equal likelihood of choosing the least favourable action as it is to select the second-best one. This approach may be deemed unsatisfactory in scenarios where the sub-optimal actions carry significant negative consequences. To address this limitation, a straightforward solution is to assign probabilities to each action based on their respective estimated value. In this way, one can expect that the greedy action retains the highest selection probability while all others are ranked and weighted according to their value estimates. These modified action selection rules fall under the category of softmax methods. In most cases, a softmax approach employs a Gibbs or Boltzmann distribution, determining the probability of selecting an action the at time step $t$ according to the probability function $P_t(a)$ defined as follows:

$$P_t(a) = \frac{e^{Q_t(a)/\tau}}{\sum_{a' \in \mathcal{A}} e^{Q_t(a')/\tau}}. \tag{3.13}$$

Here, $\tau$ represents a positive parameter known as the temperature. Elevated *temperatures* lead to actions being almost equally probable, whereas lower *temperatures* result in a more pronounced disparity in selection probability among actions with distinct value estimates. As $\tau$ approaches zero, softmax action selection converges to the same outcome as greedy action selection. It is worth mentioning that the softmax effect can be achieved through various methods beyond a Gibbs distribution. For instance, an alternative approach involves adding a random number from a long-tailed distribution to each value estimate and subsequently selecting the action with the highest cumulative sum.

### 3.2.4 *Action value estimation approaches*

Some methods are necessary for estimating the quality of an action before applying the intended exploration strategy. These methods, referred to as action-value methods, aim to estimate the expected return of an action in terms of reward when applied in the underlying environment. One intuitive and natural way to provide an estimation over time is through the so-called sample-average method, which represents the average reward obtained from performing that action:

$$Q_t(a) = \frac{\text{sum of rewards when } a \text{ is taken before } t}{\text{number of times } a \text{ is taken before } t.} \tag{3.14}$$

This represents one of the simplest methods for estimating the rewards. For example, it can be easily modified to give more importance to the most recent values or keep a memory of the previous rewards. However, this can result in computational inefficiency. Let us see an alternative efficient way to implement this action-value estimation recursively. We will indicate with $Q_n(a)$ the estimation of action $a$ after it has been selected $n - 1$ times:

$$Q_n(a) = \frac{R_1 + R_2 + \cdots + R_{n-1}}{n-1}. \tag{3.15}$$

From the above expression, one can easily notice that deriving a recursive formula for updating averages with small computation requirements to process each new reward can be easily derived, as illustrated below:

$$\begin{aligned} Q_{n+1}(a) &= \frac{1}{n} \sum_{i=1}^{n} R_i \\ &= \frac{1}{n} \left[ \sum_{i=1}^{n-1} R_i + R_n \right] \\ &= \frac{1}{n} \left[ \frac{1}{n-1}(n-1) \sum_{i=1}^{n-1} R_i + R_n \right] \\ &= \frac{1}{n} \left[ (n-1) Q_n(a) + R_n \right] \\ &= Q_n(a) + [R_n - Q_n(a)]. \end{aligned} \tag{3.16}$$

This approach greatly reduces the amount of computational resources and does not require keeping track of all the previous action values but only of its estimation obtained so far.

However, even if this estimation method is computationally efficient, it is fully suitable to deal with stationary problems. In the case when RL problems are non-stationary, a common approach consists of changing the (3.16) by introducing a constant step-size parameter $\alpha \in (0, 1]$ as follows:

$$\begin{aligned} Q_{n+1}(a) &= Q_n(a) + \alpha [R_n - Q_n(a)] \\ &= \alpha R_n + (1-\alpha) Q_n(a) \\ &= \alpha R_n + (1-\alpha) [\alpha R_{n-1} + (1-\alpha) Q_{n-1}(a)] \\ &= \cdots = \\ &= (1-\alpha)^n Q_1 + \sum_{i=1}^{n} \alpha (1-\alpha)^{n-i} R_i. \end{aligned} \tag{3.17}$$

From this expression, one can easily note how this represents a weighted version of (3.16) where more importance is given to the most recent rewards, while the importance of previous rewards gradually decreases.

Although these methods have different characteristics and advantages depending on the scenario on which they are applied, they both still hold a common aspect: they are dependent to some extent on the initial action-value estimation $Q_1(a)$, which in statistical language, this means that these methods are *biased* by their initial estimates. This usually does not represent a problem, but sometimes, the initial value needs to be set carefully since, through sample-average methods, the bias disappears once all actions have been selected at least once. On the other hand, for methods with constant $\alpha$, the bias is permanent, though decreasing over time. Usually, a common practice is to set the initial values to zero.

## 3.3 Dynamic programming

Dynamic program (DP) represents a set of algorithms used to learn the optimal policy in RL problems that can be modelled as MDPs. Their main characteristic lies in the fact that they assume a perfect model of the environment. For this reason, their high computational requirements limit their applications. In addition, although they can be theoretically applied to problems with continuous state–action space, in practice, they can only provide approximated solutions to these cases, keeping their full suitability to problems where the respective environments are represented through finite MDP, i.e. state, action, and reward sets are finite. The dynamics are represented through a set of probabilities $\mathscr{P}(s', r|s, a)$ for all $s \in \mathscr{S}$ and $a \in \mathscr{A}$. However, they are worth illustrating, as they provide an essential foundation for understanding the majority of RL methods discussed later in this book.

Under the aforementioned assumptions and on the relationship between the $Q$ function and $V$ value function defined in Section 3.1.2, using the Bellman equation, we can express the optimal value function recursively as follows:

$$V^*(s) = \max_{a \in \mathscr{A}} \mathbb{E}\left[\mathscr{R}_{t+1} + \gamma V^*(\mathscr{S}_{t+1}) \mid \mathscr{S}_t = s, \mathscr{A}_t = a\right]$$
$$= \max_{a \in \mathscr{A}} \sum_{s', r} \mathscr{P}(s', r|s, a)\left[r + \gamma V^*(s')\right]. \tag{3.18}$$

The reader can easily derive the same for the $Q$-value function. However, DP mainly focuses on the $V$-value function. Starting from that, we first need to consider a possible way to compute the $V$-value function for an arbitrary policy $\pi \in \Pi$, referred to as **policy evaluation** in the DP literature. Introducing the term $G_t = \sum_{k=0}^{\infty} \gamma^k r_{t+k+1}$, and using Bellman's equation, for all $s \in \mathscr{S}$ we can write:

$$\begin{aligned} V_\pi(s) &= \mathbb{E}_\pi\left[G_t \mid \mathscr{S}_t = s\right] \\ &= \mathbb{E}_\pi\left[R_{t+1} + \gamma G_{t+1} \mid \mathscr{S}_t = s\right] \\ &= \mathbb{E}_\pi\left[R_{t+1} + \gamma V_\pi(\mathscr{S}_{t+1}) \mid \mathscr{S}_t = s\right] \\ &= \sum_a \pi(a|S) \sum_{s', r} \mathscr{P}(s', r|s, a)\left[r + \gamma V_\pi(s')\right], \end{aligned} \tag{3.19}$$

in which $\pi(a|s)$ represent the probability of taking an action $a$ conditioned by the fact of being in the state $s$ and under the policy $\pi$.

From (3.19), one can easily notice that, if the dynamics are completely known, i.e. state transition probabilities, this represents a system of $\|\mathscr{S}\|$ linear equation. However, a straight calculation can be tedious. A common way to solve this policy estimation problem is through an iterative approach called **iterative policy evaluation**. This methods uses Bellman's equation to provide a successive approximation of the optimal $V$-value function, starting from an arbitrary initial policy $V_0$, as follows:

$$\begin{aligned} V_{k+1} &= \mathbb{E}_\pi\left[\mathscr{R}_{t+1} + \gamma V_k(\mathscr{S}_{t+1}) \mid \mathscr{S}_t = s\right] \\ &= \sum_a \pi(a|s) \sum_{s', r} \mathscr{P}(s', r|s, a)\left[r + \gamma V_k(s')\right], \end{aligned} \tag{3.20}$$

---
**Algorithm 3.1** Iterative policy evaluation, for estimating $V \approx V_\pi$
---
**Require:** Initialise the initial value of $V(s)\ \forall s \in \mathscr{S}$; a small threshold $\theta > 0$ determines the accuracy of estimation.
1: **while** $\Delta > \theta$ **do**
2:     **for** $s \in \mathscr{S}$ **do**
3:        $v \leftarrow V_k(s)$
4:        $V_{k+1}(s) \leftarrow \sum_a \pi(a|s) \sum_{s',r} \mathscr{P}(s',r|s,a)\left[r + \gamma V_k(s')\right]$
5:        $\Delta \leftarrow |v - V_{k+1}(s)|$
6:     **end for**
7: **end while**
---

for all $s \in \mathscr{S}$. Such iterative evaluation, which can be shown to converge to a policy $\pi \in \Pi$ as $k \to \infty$, substitutes the previous value assigned to $s$ with a fresh value derived from the prior values of the successor states of $s$, as well as the anticipated immediate rewards, encompassing all potential one-step transitions under the policy undergoing evaluation. A complete pseudocode for iterative is illustrated in Algorithm 3.1.

Following this algorithm will allow us to find specific policies. However, the final goal of a RL-based algorithm is to find the optimal policy $\pi^*$. Indeed, based on Algorithm 3.1, we know how good it is to follow the estimated policy, but the question is if it would be better or worse to change to the new policy and see if selecting another action in a particular state $s$ would bring better performances. One way to answer this question is to evaluate the quality of an action $a$ when in the state $s$ and check whether this is greater than or less than $V_\pi(s)$. This is performed by using the $Q$-value function for that action previously defined. To illustrate better this aspect, let us consider a pair of deterministic policies $\pi(s)$ and $\pi'(s)$. Then we want to check if for:

$$Q_\pi(s, \pi'(s)) \geq V_\pi(s), \tag{3.21}$$

for all $s \in \mathscr{S}$. If this happens, this means that the policy $\pi'$ must be as good as, or better than $\pi$, and that:

$$V_{\pi'}(s) \geq V_\pi. \tag{3.22}$$

The relationship in (3.22) is proven through the policy improvement theorem obtained through the set of equations illustrated in [3], which are obtained by recursively applying (3.21) and expanding the through Bellman's equation as follows:

$$\begin{aligned} V_\pi(s) &\leq Q_\pi(s, \pi'(s)) \\ &= \mathbb{E}\left[\mathscr{R}_{t+1} + \gamma V_\pi(S_{t+1}) | S_t = s, A_t = \pi'(s)\right] \\ &= \mathbb{E}_{\pi'}\left[\mathscr{R}_{t+1} + \gamma V_\pi(S_{t+1}) | S_t = s\right] \\ &\leq \mathbb{E}_{\pi'}\left[\mathscr{R}_{t+1} + \gamma Q_\pi(S_{t+1}, \pi'(S_{t+1})) | S_t = s\right] \\ &= \mathbb{E}_{\pi'}\left[\mathscr{R}_{t+1} + \gamma \mathbb{E}\left[\mathscr{R}_{t+2} + \gamma V_\pi(S_{t+2}) | S_{t+1} = s, A_{t+1} = \pi'(S_{t+1})\right] | S_t = s\right] \\ &= \cdots \\ &\leq \mathbb{E}\left[\mathscr{R}_{t+1} + \gamma \mathscr{R}_{t+2} + \gamma^2 \mathscr{R}_{t+3} + \cdots\right] \\ &= V_{\pi'}(s). \end{aligned} \tag{3.23}$$

In summary, considering all the possible states and the improvement theorem, we can express the new improved policy $\pi'$ from the original policy $\pi$ as

$$\begin{aligned}
\pi'(s) &= \arg\max_{a \in \mathcal{A}} Q_\pi(s, a) \\
&= \arg\max_{a \in \mathcal{A}} \mathbb{E}\left[\mathcal{R}_{t+1} + \gamma V_\pi(S_{t+1}) | S_t = s, A_t = a\right] \\
&= \arg\max_{a \in \mathcal{A}} \sum_{s', r} \mathcal{P}(s', r | s, a) \left[r + \gamma V_\pi(s')\right],
\end{aligned} \quad (3.24)$$

which means that the $V$-value function for the new policy is expressed as

$$V_{\pi'}(s) = \max_{a \in \mathcal{A}} \sum_{s', r} \mathcal{P}(s', r | s, a) \left[r + \gamma V_{\pi'}(s')\right]. \quad (3.25)$$

Such improvement process can be repeated continuously until we reach the optimal policy, which is assumed to exist because a finite MDP has only a finite number of deterministic policies. This process, illustrated through Algorithm 3.2, is referred to as **policy iteration**, which involves the iterative evaluate-improve-evaluate process of a policy.

However, this process has the main drawback that it can result in computational inefficiency since it will require several sweeps through the state set. To address

---

**Algorithm 3.2** Policy iteration for estimating $\pi \approx \pi*$

**Require:** Initialise the initial value of $V(s)\ \forall s \in \mathcal{S}$; a small threshold $\theta > 0$ determines the accuracy of estimation.

1: **Policy Evaluation**
2: **while** $\Delta > \theta$ **do**
3:     **for** $s \in \mathcal{S}$ **do**
4:         $v \leftarrow V(s)$
5:         $V(s) \leftarrow \sum_{s', r} \mathcal{P}(s', r | s, a) [r + \gamma V_k(s')]$
6:         $\Delta \leftarrow |v - V(s)|$
7:     **end for**
8: **end while**
9: **Policy Improvement** stablePolicy $\leftarrow$ TRUE
10: **for** $s \in \mathcal{S}$ **do**
11:     actionOld $\leftarrow \pi(s)$
12:     $\pi(s) \leftarrow \arg\max_{a \in \mathcal{A}} \sum_{s', r} \mathcal{P}(s', r | s, a) [r + \gamma V_\pi(s')]$
13:     **if** actionOld $\neq \pi(s)$ **then**
14:         stablePolicy $\leftarrow$ FALSE
15:     **end if**
16: **end for**
17: **if** stablePolicy == TRUE **then**
18:     End Evaluation;
19: **else**
20:     Continue Evaluation and then Improvement
21: **end if**

---
**Algorithm 3.3** Value iteration, for estimating $\pi \approx \pi^*$
---
**Require:** Initialise the initial value of $V(s) \, \forall s \in \mathscr{S}$; a small threshold $\theta > 0$ determines the accuracy of estimation.
1: **while** $\Delta > \theta$ **do**
2:      **for** $s \in \mathscr{S}$ **do**
3:          $v \leftarrow V_k(s)$
4:          $V_{k+1}(s) \leftarrow \max_{a \in \mathscr{A}} \sum_{s',r} \mathscr{P}(s',r|s,a)\left[r + \gamma V_k(s')\right]$
5:          $\Delta \leftarrow |v - V_{k+1}(s)|$
6:      **end for**
7: **end while**
---

this, exploiting the fact that policy iteration can be truncated in several ways without losing the convergence, a more efficient computational way to find the optimal policy is through the **value iteration** technique that combines the policy improvement and truncated policy evaluation as illustrated in Algorithm 3.3.

In summary, independently from the type of adopted algorithm, the process is still the same, which is usually referred a general policy iteration (GPI). This represents the process in which policy evaluation and policy improvement alternate to converge to the optimal policy. Indeed, if both processes stabilise, the value function and policy must be optimal.

## 3.4 Monte Carlo methods

As mentioned at the beginning of the previous section, the main assumption for the use of DP was the complete knowledge of the environment, which in most cases is either partly known or completely unknown. In such cases, the use of Monte Carlo results very helpful. Indeed, the estimation of either the $V$-value function or $Q$-value through these methods is based on the experience obtained from actual or simulated interaction with an environment, i.e. sample sets containing sequences of states, actions, and rewards. Then, one can easily notice how only minimal knowledge about the environment, such as all the possible transitions from one state to another, is necessary to be known to generate the experience samples.

What Monte Carlo methods do in solving RL problems is to average the sample return of each experience. In this context, we say that experience is gathered from the *episodes*. In other words, the experience is divided into episodes with a start and an end, and then $V/Q$ values are estimated and the respective policy changed. From this, one can also note another difference with the DP approach: Monte Carlo methods are incremental in an episode-by-episode sense but not in a step-by-step, i.e. *online* sense like DP. However, even with this substantial difference with DP methods, the idea behind GPI for obtaining the optimal policy after evaluation can still be applied.

## 3.4.1 Estimation procedures

Let us first provide a brief introduction on how a Monte Carlo method works to estimate the state-value function. A very intuitive way of doing this is by averaging the rewards observed after visiting the interested state. Suppose that we want to estimate $V_\pi(s)$ for a particular state $s$ and under a generic policy $\pi$, given a set of experience episodes, in which each occurrence of state $s$ within an episode is called a **visit** to $s$. For each episode, the specific state $s$ may be visited multiple times. For this reason, two different approaches of the Monte Carlo methods are defined as

- **First-visit** estimating the state-value function as the average of the returns following first visits to $s$;
- **Every-visit** estimating the state-value function as the average of the returns following all visits to $s$.

Although it is proven that according to the law of large numbers, the sequence of averages of these estimates converges to their expected value, we will consider the first-visit approach, which is represented in Algorithm 3.4. Let us consider an example in which we get experience from a system with three possible states named $S_A$, $S_B$, and $S_T$, where the last one represents the termination state. We assume to receive the two following episodes:

$$\begin{aligned}\text{Episode 1: } & S_A, +3 \rightarrow S_A, +2, \rightarrow S_B, -4 \rightarrow S_A, +4, \rightarrow S_B, -2, \rightarrow S_T; \\ \text{Episode 2: } & S_B, -2 \rightarrow S_A, +3 \rightarrow S_B, -3,\end{aligned} \quad (3.26)$$

in which the notation $S, R \rightarrow S'$ indicate that the system visit state $S$ where an action is performed, receiving a reward $R$, and move to state $S'$.

For the sake of simplicity, we suppose $\gamma = 1$. According to the first-visit approach, the estimation of the two states based on the received episodes are as follows:

$$\begin{aligned} V_\pi(S_A) &= ((3+2-4+4-2) + (3-3))/2 = 1.5; \\ V_\pi(S_B) &= ((-4+4-2) + (-2+3-3))/2 = -1 \end{aligned} \quad (3.27)$$

This type of estimation can also be applied to the action-value function. Indeed, sometimes it is necessary to estimate the $Q_\pi(s, a)$ function, especially when we do

---

**Algorithm 3.4** First-visit Monte Carlo state-value estimation

**Require:** A policy $\pi$ to evaluate; A set $N$ of episodes.
1: **for** each episode **do** Consider the $i$-th episode following the generic policy $\pi$ with a temporal index $EP_i^\pi : S_0, R_0 \rightarrow S_1, R_1 \rightarrow S_2, R_2 \rightarrow S_3, R_3 \rightarrow, \cdots$
2:     **for** each time step of the considered $i$-th episode **do**
3:         **for** each state S **do** $V_i(s) \leftarrow \gamma V_i(s) + R_{t+1}$
4:         **end for**
5:     **end for**
6: **end for**
7: **return** $V^\pi(s) = \frac{1}{N} \sum_{i=1}^{N} V_i(s)$

**Algorithm 3.5** GPI for Monte Carlo methods

**Require:** A random policy $\pi$; a set $N$ of episodes.
1: **for** each episode **do**
2:     **for** each step of the episode **do**
3:         Evaluate the $Q(s, a)$ function.
4:         Improve the obtained policy
5:     **end for**
6: **end for**
7: **return** The optimal obtained policy

not have complete information about the model to simulate. The Monte Carlo methods for this type of function estimation are essentially the same where a state–action $(s, a)$ pair is said to be visited in an episode if the state $s$ is visited and action $a$ is taken in it. As a result, the first-visit Monte Carlo method is an estimation method that computes the average returns by considering the outcomes following the initial occurrence of state visitation and the action selection in each episode. A corresponding algorithm can be easily obtained from the one illustrated in Algorithm 3.4. However, in this case, there is one aspect that needs to be carefully addressed. Indeed, during episode generation, there is always the risk that many state–action pairs may never be visited. Then, it is necessary to put in place mechanisms that assure continuous exploration, for example, by specifying that in all episodes, every pair has a non-zero probability of being selected as the start.

### 3.4.2 General policy iteration in Monte Carlo methods

As already anticipated and explained for the DP method, the GPI can also be applied to Monte Carlo methods. This means that also in this case we constantly have an alternation between value function evaluation and policy estimation and policy improvement. More specifically, the value function undergoes iterative adjustments to closely approximate the value function corresponding to the current policy. Simultaneously, the policy undergoes repeated enhancements based on the current value function.

Supposing that we do not have complete knowledge about the model, we estimate the state–action value function. In this case, the evaluation process is done through the generation of episodes and the calculation of the average function over all the episodes. The improvement is performed by deterministically selecting for each state $s$ the action $a$, maximising the function, and obtaining the policy for the next iteration as:

$$\pi_{k+1}(s) = \arg\max_{a} Q_k^{\pi}(s, a). \tag{3.28}$$

Also, in this case, we notice that this approach can result in computationally inefficiency since we need to evaluate all the episodes before the evaluation and then perform the improvement. A common approach usually adopted is to perform

**Algorithm 3.6** GPI for $\varepsilon$-soft Monte Carlo methods
***
**Require:** A random policy $\pi$; a set $N$ of episodes.
1: **for** each episode **do**
2:     **for** each step of the episode **do**
3:         Evaluate the $Q(s, a)$ function.
4:         Find the optimal action $A^* = \arg\max_a Q(s, a)$
5:         Update the probabilities according to (3.29).
6:     **end for**
7: **end for**
8: **return** The optimal obtained policy
***

the evaluation-improvement mechanisms within each episode as illustrated through Algorithm 3.6.

Together, these mechanisms contribute to approaching the optimal value. It is worth mentioning that, in addition to being more computationally efficient, such algorithm also permits a more rapid convergence, making the assumption of an infinite number of episodes unnecessary. However, some space–action pairs might not be explored during the process. To address this, two main methods can be employed:

- **On-policy** methods: These methods try to directly improve the policy used to make decisions.
- **Off-policy** methods: These methods are intended to improve the way in which the data for episodes is generated.

Explaining these types of policies is out of the scope of this book, and the interested reader can find more information in [4]. However, an example of the on-policy is illustrated to provide a brief introduction to this exploration issue. In particular, a very intuitive way of implementing an on-policy is by adopting the $\varepsilon$-greedy approach. This consists of slightly changing the policy's improvement process by assigning probabilities of taking an action in a particular state. More specifically, indicating with $A^*$ the optimal action that maximises the state–action value for a particular state $S$, the probability of executing any other action is updated as follows:

$$\pi(a|S) = \begin{cases} 1 - \varepsilon + \varepsilon/|\mathscr{A}(S)| & \text{if } a = A^*, \\ \varepsilon/|\mathscr{A}(S)| & \text{otherwise,} \end{cases} \quad (3.29)$$

where $\mathscr{A}(S)$ represents the set of all the possible actions. This approach is referred to as $\varepsilon$-soft policy, for which the algorithm is illustrated below. For convergence purposes, the value of $\varepsilon$ is chosen as the inverse of the time step.

## 3.5 Temporal difference learning

Temporal difference (TD) learning is a type of RL algorithm that combines aspects of DP and Monte Carlo methods for estimating either the $V$-value function or the $Q$-value function of an MDP. For this reason, it represents a novel central idea in the area of RL. Indeed, as with Monte Carlo methods, TD learning can learn directly from raw experience without having knowledge/model of the environment, while, like DP, they are able to estimate the desired function without waiting for a final outcome. In addition, this type of approach is also powerful in dealing with non-stationary processes.

### 3.5.1 Estimation procedures

TD learning can be applied to learning state-value functions ($V$-function) and state–action value functions ($Q$-function). The choice depends on the specific problem and the goal of the RL task. Then, without loss of generality, we will focus on the case of $V$-value function estimation for a given policy $\pi$. In this case, the usage of TD learning approach for estimating the value of a particular state at time $s$ ($s_t$) is expressed as

$$V(s_t) = V(s_t) + \alpha \left[ R_{t+1} + \gamma V(S_{t+1}) - V(S_t) \right]. \tag{3.30}$$

This means that while in the state $S_t$, we take an action, observe a reward $R_{t+1}$, to which state we transition $S_{t+1}$, and then update its value. It is worth mentioning that the quantity between the square bracket is commonly referred to as TD error $\delta_t = R_{t+1} + \gamma V(S_{t+1}) - V(S_t)$, which is related to the estimation rule for non-stationary problems illustrated in (3.17).

Based on (3.30), we can then say that TD learning methods update their estimates based in part on other estimates, i.e. they learn a guess from a guess. This type of estimation working rule has clear advantages compared to DP approaches since a model of the environment, its reward and next-state probability distributions are not required. On the other hand, compared to Monte Carlo methods, TD learning can be naturally implemented in an online incremental fashion without waiting for the end of an episode. However, the main question that can arise from using this approach is whether it still guarantees convergence of the estimation. In that case, it has been proven that TD methods converge asymptotically to the correct predictions, as with Monte Carlo methods.

### 3.5.2 General policy iteration in temporal difference learning: on-policy and off-policy

Now that we have a clear understanding of how TD methods estimate the value function of interest, let us see how the GPI for policy estimation is performed. As for the Monte Carlo methods, we may face the need to trade-off between exploration and exploitation, leading to on-policy and off-policy approaches depending on where the improvement of the policy is performed. To illustrate both approaches, let us consider the case of improving the policy based on the estimation of the $Q$-value function.

Within a TD learning on-policy, the value of the state–action function is performed as follows:

$$Q(s_t, a_t) = Q(s_t, a_t) + \alpha \left[ R_{t+1} + \gamma Q(s_{t+1}, a_{t+1}) - Q(s_t, a_t) \right]. \tag{3.31}$$

Once the function is estimated, the correspondent policy can be improved by using either the $\varepsilon$-greedy or $\varepsilon$-soft policy for assigning the action probabilities. Also, in this case, for convergence assurance, the value of $\varepsilon$ is chosen as the inverse of the time step.

On the other hand, the off-policy approach, usually referred to as **Q-learning**, directly approximates the optimal policy on the estimation process, which in this case is changed as follows:

$$Q(s_t, a_t) = Q(s_t, a_t) + \alpha \left[ R_{t+1} + \gamma \max_a Q(s_{t+1}, a) - Q(s_t, a_t) \right]. \tag{3.32}$$

In other words, this approach dramatically simplifies the analysis of the algorithm by directly estimating the optimal action-value function independently from the policy being followed. The policy still has an effect – it determines which state–action pairs are visited and updated.

## 3.6 Summary

This chapter has briefly introduced the MDP and how it is related to the RL problems. It has also given an essential understanding of the multi-arm bandit problem and how the exploration vs exploitation dilemma is addressed in estimating either the $V$-value or $Q$-value function. Finally, an introduction to the most common methods for optimal policy estimation, such as DP, Monte Carlo, and TD, has also been provided.

## References

[1] Sutton RS and Barto AG *Introduction to Reinforcement Learning*. vol. 135. MIT Press Cambridge; 1998.
[2] Puterman ML. *Markov Decision Processes: Discrete Stochastic Dynamic Programming*. John Wiley & Sons, Inc.; 1994.
[3] Jaakkola T, Singh S, and Jordan M. Reinforcement learning algorithm for partially observable Markov decision problems. In: Tesauro G, Touretzky D, and Leen T, editors. *Advances in Neural Information Processing Systems*. vol. 7. MIT Press; 1994.
[4] V. François-Lavet, P. Henderson, R. Islam, M. G. Bellemare, and J. Pineau, An Introduction to Deep Reinforcement Learning. now Publishers Inc., 2018.

*Chapter 4*
# Value function approximation for continuous state–action space

The previous chapter illustrates methods for estimating the value of either $V$-value or $Q$-value function and how those results are helpful for the estimation of the optimal policy. However, in most real-life applications, the state–action space can be extremely and arbitrarily large, making it extremely difficult to apply these methods. We cannot expect to find an optimal policy or the optimal value function even within the limits of infinite time and data. Furthermore, the aforementioned methods store the state–action mapping into tables, meaning that large spaces require large storage capabilities. To tackle these issues, this chapter introduces the approximation methods used to find good approximated solutions of the value function of interest with limited computational capabilities. Such methods represent the start towards the implementation of DRL-based methods.

## 4.1 From tabular to function approximation method

Although the methods introduced in the previous chapter mentioned *functions* such as the $V$-value or $Q$-value function used to find the optimal policy, these must be considered as tabular methods for approximating the optimal policy. In other words, we can imagine the agent filling a table with the best action to perform for each state. However, as the size of the problem increases, one can easily imagine how this approach becomes unfeasible. Then, alternative methods are necessary.

As the size of the problem becomes continuous, the most intuitive method to adopt is to represent the value function of interest through a parameterised function, where the parameters, also referred to as weights, are represented through a vector $(w) \in \mathbb{R}^d$. Then, for example, the $V$-value function will be written as $V(s, \mathbf{w})$. It can be a linear function of the feature of the state weighted according to $\mathbf{w}$, or a more general complex function. In this case, the main objective for estimating that function will be to find the optimal value for the weights. Usually, the number of weights, i.e. the dimensionality $d$ of $\mathbb{R}^d$, results in less than the number of states, and changing one weight changes the estimated value of many states. Furthermore, since RL is applied to very complex problems, as already mentioned in Chapter 2, these parametric functions are usually represented by DNN.

In summary, representing the value function through a parameterised function results in the problem of finding the optimal weight which approximates the function. For this reason, the subsequent part of the chapter illustrates some of the more common algorithms for the estimation of the weights. These algorithms dynamically adjust the weights along several iterations. To estimate the value of the optimal parameters, i.e. weights of the parameterised function, it is necessary to define an objective function for prediction. Without loss of generality, referring to the $V$-value function, the objective function for its approximation is defined as follows:

$$L(\mathbf{w}) = \mathbb{E}\left[(V_\pi(s) - V(s, \mathbf{w}))^2\right]. \tag{4.1}$$

This is usually referred to as the loss function, which gives a rough measure of how much the approximated value function differs from the true value. In addition, it is important to keep in mind that $V_\pi(s)$ is referred to as the target. This will be helpful later when talking about deep Q-learning.

The main idea behind using this objective function is to find the optimal vector $\mathbf{w}^*$ for which $L(\mathbf{w}^*) \leq L(\mathbf{w}); \ \forall \mathbf{w} \in \mathbb{R}^d$. Reaching this point of global optimum is possible for simple approximation functions such as linear functions, while it is rarely possible for complex approximating functions like the ones implemented through DNN. In the latter cases, the optimisation algorithm is not guaranteed to converge to a global minimum but to local minima where the condition $L(\mathbf{w}^*) \leq L(\mathbf{w})$ is valid in the neighbourhood of $\mathbf{w}^*$.

The most widely used method for parameter updates is the stochastic gradient descent. In particular, indicating with $t$ the temporal evolution within an episode, the value of the weights for the next value *evaluation* are updated as follows:

$$\mathbf{w}_{t+1} = \mathbf{w}_t + \alpha \left[V_\pi(S_t) - V(S_t, \mathbf{w}_t)\right] \nabla V(S_t, \mathbf{w}_t). \tag{4.2}$$

It is worth mentioning that the error is propagated back through backpropagation if we use a multi-layer NN as the approximator of the value function. Details about backpropagation have been provided in Chapter 2.

## 4.2 Deep Q-learning

As its name already suggests, deep Q-learning represents the combination of DNN and Q-learning to find the optimal policy for the RL agent. The principle is still the same: mapping each state–action pair within a value. However, the main difference is that while Q-learning aims at obtaining a sort of table, deep Q-learning (DQL) aims to find the $Q$-value function for each possible action in each state. The main difference between the two methods is summarised in Figure 4.1. From this Figure, one can easily notice that a DQL approach the algorithm, for each state provided as input, will provide the $Q$-value for each possible action. If we adopt a greedy policy, we will select the action for which the DNN provide the maximum value.

Now that we have an idea about the DQL, the next step would be to understand how to train such a network in an RL fashion, i.e. through action taken in the environment, received reward and next-state transition. First of all, we need to define the

Value function-based RL   55

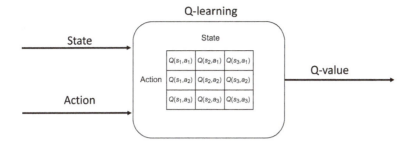

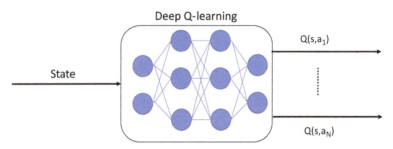

*Figure 4.1  Difference between Q-learning (top) and deep Q-learning (bottom) approach*

loss function and then the target function to calculate the loss. In this case, the target function is the one obtained using Bellman's equation, which for a generic tuple $\langle s, a, r, s' \rangle$ in the $i$-th episode, we indicate as follows:

$$Q_\pi(s, a) = R_i + \gamma \max_{a' \in \mathcal{A}} Q(s', a'). \tag{4.3}$$

The loss function assumes the form of

$$L(\mathbf{w}) = (R_i + \gamma \max_{a' \in \mathcal{A}} Q(s', a') - Q(s, a, \mathbf{w}))^2. \tag{4.4}$$

The main questions arising from (4.4) are (i) how the target function is implemented and (ii) how, in practice, the DNN representing the Q-function in Figure 4.1 is trained. The answer to the first question is quite simple. Indeed, the $Q$-value for the target function is generated by using another DNN with the same network that we want to train. However, there is a particular aspect of the weight of each function that we will discuss later. With regard to the latter question, the DQL algorithm is trained by using an experience replay buffer. Such buffer, whose content is continually updated as the agent interacts with the environment, consists of storing a finite number of experience samples $\langle s, a, r, s' \rangle$ in a buffer, which is randomly sampled to create a mini-batch used to perform the updates of the weights, as below:

$$\mathbf{w}_{t+1} = \mathbf{w}_t + \alpha \left[ R_i + \gamma \max_{a' \in \mathcal{A}} Q(s', a') - Q(s, a, \mathbf{w}) \right] \nabla_\mathbf{w} Q(s, a, \mathbf{w}), \tag{4.5}$$

in which $\alpha$ represents the learning rate. The update rule expressed in (4.5) is the one according to the SGD, in which the update of the network weights is performed after calculating the loss function for each sample stored in the batch. However, the error update can be performed by using any other method illustrated in Chapter 2. In addition, for the batch provided as training set, it is worth mentioning that the experience samples used for training the $Q$ network are not provided in the correct temporal sequence. Instead, the experience replay is randomly sampled in a way that allows the mini-batch to provide non-correlated examples. This approach of breaking the temporal correlation leads to more efficient learning with an optimal bias-variance trade-off, i.e. the resulting optimal policy will show optimal performances in general contexts.

With regard to the loss, as mentioned before, two slightly different networks are used, namely the main $Q$ network and the target $Q$ network. These networks have the same architecture but with different weight values. More specifically, the weight of the two networks is in a temporal gap relationship. More specifically, indicating with $\mathbf{w}^-$ a past weight configuration in the main $Q$ network, the loss function in (4.4) assumes the following form:

$$L(\mathbf{w}) = (R_i + \gamma \max_{a' \in \mathscr{A}} Q(s', a', \mathbf{w}^-) - Q(s, a, \mathbf{w}))^2. \tag{4.6}$$

Using a separate target network helps improve the stability of the algorithm since it adds a delay between the two functions. However, the $\mathbf{w}^-$ values are also updated within the training phase. In particular, the frequency with which the weights of the target network are updated is a hyperparameter, usually referred to as either **target update rate** or **target network update interval**, which can be adjusted based on the specific requirements of the problem. Usually, the updates are performed by following a **fixed interval update**, in which the updates occur every $C$ training step by copying the parameters of the main $Q$ network into the target network. However, instead of performing a hard copy of the parameters a *soft* approach can be used in which the new parameters copies in the target network will be $\mathbf{w}^- = \tau \cdot \mathbf{w} + (1 - \tau) \cdot \mathbf{w}^-$, where $\tau \in [0, 1]$ represents another hyperparameter. Then, the choice of the target update rate depends on factors such as the characteristics of the environment, as well as on the desired trade-off between stability and learning speed. For the sake of completeness and good understanding, the entire process of DQL is summarised in Algorithm 4.1.

## 4.3 Methods for sampling the replay buffer

Together with the update of the target network, the management of the experience replay buffer plays a pivotal role during the training/learning phase of DQL-based systems, especially in guaranteeing efficient and stable learning. Such buffer stores previous experience of the agent interacting with the environment in terms of performed action and received reward. Since such buffer has a limited capacity, the oldest experiences are replaced with new ones when the buffer reaches its capacity.

The most straightforward way for doing this operation is by performing a **uniform random sampling**, according to which each experience tuple has an equal

## Algorithm 4.1 Deep Q-learning algorithm

**Require:** Initialisation of the following:
1: Replay memory capacity;
2: Weights of both main and target $Q$ networks;
3: **for** each episode **do**
4:     Initialise the starting state
5:     **for** each step of the episode **do**
6:         select an action according to $\varepsilon$-greedy or softmax selection.
7:         Observe the reward and the next state.
8:         Store the observed $\langle s, a, r, s' \rangle$ tuple in the replay buffer.
9:         Randomly sample the replay buffer to create the mini-batch.
10:        Calculate the loss function as in (4.6) by using both networks.
11:        Update the $\mathbf{w}$ in the main $Q$ network.
12:        **if** $\mathbf{w}^-$ s to be updated **then**
13:            update target network weights.
14:        **end if**
15:    **end for**
16: **end for**
17: **return** The optimal obtained policy

chance of being selected. This leads to a fair representation of the entire dataset as well as contributes to breaking the temporal correlation between consecutive experiences. This approach is easy to implement, but may be inefficient in situations where certain experiences are more informative or critical for learning.

The potential inefficiency of performing a uniform sampling can be addressed through the usage of sampling methods which keep track of *most relevant* experiences. This is usually done by adopting a **prioritised experience replay** sampling. Indeed, in addition to breaking the temporal correlation, this method also assigns priorities to learning samples according to the experience loss. Typically, the priority of the $i$-th example in the buffer is directly represented through the loss function of the experience example as follows:

$$\text{Priority}(i) = L^i(\mathbf{w}) + \varepsilon, \tag{4.7}$$

where $\varepsilon$ represents a small positive value. Once these quantities are computed for each example, the probability $P(i)$ of sampling the event is:

$$P(i) = \frac{\text{Priority}(i)^\alpha}{\sum_j \text{Priority}(j)^\alpha}, \tag{4.8}$$

with $\alpha$ being a hyperparameter known as the exponent or temperature parameter. However, some other approaches can be used to calculate the priority-base probabilities, such as the softmax function. From this approach, one can easily notice that higher probabilities are assigned to experiences with higher loss, meaning that the algorithm focuses more on learning from experiences that are more challenging, which in turn will potentially lead to accelerating the learning process. In addition, priorities of experiences can also be adjusted during training. For example, experiences that were initially of high priority may become less critical over time, and vice versa, ensuring that the prioritisation aligns with the changing importance of experiences. However, priorities need to be carefully managed to avoid overemphasising individual experiences or, even worse, causing an overfitting. For these reasons, **hybrid** sampling methods that combine uniform random sampling with prioritised sampling are sometimes adopted. By doing so, they leverage the benefits of both methods.

## 4.4 Double DQL

While talking about the methods for sampling the replay buffer to generate the batch for training, it has already been mentioned that DQL can suffer from overestimation bias. This brought about the concept of double DQL (DDQL). Indeed, DDQL represents a pure extension of traditional deep Q-learning that aims to address overestimation bias by employing the main Q-network network for action selection and the target network for action evaluation but in a different way [1]. In particular, indicating the target network as $Q_T$ and the main network as $Q_M$, the loss function for the generic $i$-th experience example in the buffer is calculated as follows:

$$L^i(\mathbf{w}) = (R_i + \gamma \cdot Q_T(s', \arg\max_{a' \in \mathcal{A}} Q_M(s, a, \mathbf{w}), \mathbf{w}^-) - Q_M(s, a, \mathbf{w}))^2. \quad (4.9)$$

From (4.9), one can easily notice the concept of employing two different networks, one for action selection and one for action evaluation. Decoupling these two operations through two different networks ensures, first of all, more stability in the training. In addition, having a dual-network architecture addresses the overestimation bias inherent in traditional DQL. It can be said that, compared to traditional DQL, DDQL offers a more nuanced and effective approach to RL in challenging and dynamic environments.

## 4.5 Dueling DQL

The concept of dueling DQL represents another important example of value-based RL methods. Such approach proposed in [2] is based on the concept of the advantage function, which, as was already explained in Chapter 3, is defined as follows:

$$A(s, a) = Q(s, a) - V(s). \quad (4.10)$$

This function represents the goodness of taking the action $a$ compared to the expected return in the state $s$ when following the policy $\pi$. More specifically, the dueling DQN uses the expression in (4.10) to decompose the Q-function into two streams, the value and advantage functions, which are combined together via an aggregation layer and share a common structure used to represent the state. The main benefit of this approach is that it provides the possibility to learn the most valuable states as well as to learn the most valuable action per state, avoiding exploring all the possible actions *unnecessary*, and improving the convergence speed. Indeed, it is of paramount importance to know which action to take in some cases, while in many others, the choice of action has no influence. For example, imagine training an RL-based algorithm to find the optimal path for a drone within an urban area containing buildings of different heights. In some situations, certain movements must be given more priority to avoid collisions.

According to its definition, the architecture of a dueling network is provided in Figure 4.2. The input data is first passed through an input layer, typically a convolutional NN, used to represent the state $s$ of the system. Subsequently, state representation is split into two parallel identical flows, representing the input for the state value function and the advantage value function, respectively. Finally, the outputs from the network are aggregated together through an aggregation layer, producing the $Q$-value function. In other words, such dueling network architecture can be viewed as a single $Q$-network with two streams, enabling the possibility of learning which states are or are not valuable without having to learn the effect of each action for each state. On the other hand, since the final output is still the $Q$-value, is it possible to train such a network using one of the algorithms illustrated so far in previous sections?

Let us now have a closer look at the design of such a network architecture. To this end, we will indicate with $\Psi$ the coefficient vector of the input layer, while with $\Lambda$ and $\Phi$, the coefficients of the $V$-value and $A$-value function networks, respectively. Then, we can write the mathematical model for a dueling network as follows:

$$Q(s, a, \Psi, \Lambda, \Phi) = V(s, \Psi, \Lambda) + A(s, a, \Psi, \Phi), \tag{4.11}$$

where the value outputted from $V(s, \Psi, \Lambda)$ needs to be replicated $|\mathscr{A}|$ times, i.e. the number of possible actions. From (4.11), it is clear that since the $Q$-value is obtained through a parameterised function, the result does derive a unique expression for both the $V$-value and $A$-value functions, leading to poor performance. This problem, usually referred to as the identifiability problem, is solved by forcing the

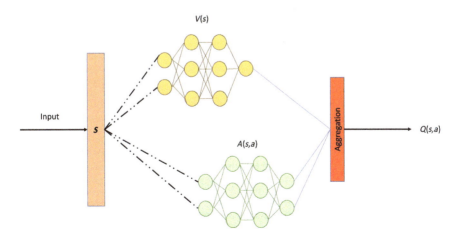

*Figure 4.2* Dueling DQN architecture with value function (top layer) and advantage function (bottom layer)

advantage function to get the zero value when the best action is chosen. In doing so, the aggregation layer is implemented as

$$Q(s, a, \Psi, \Lambda, \Phi) = V(s, \Psi, \Lambda) + (A(s, a, \Psi, \Phi) - \max_{a' \in \mathscr{A}} A(s, a', \Psi, \Phi)). \quad (4.12)$$

It can be easily noticed that when the optimal action is chosen, i.e. $a^* = \arg\max_{a' \in \mathscr{A}} A(s, a', \Psi, \Phi)$ we obtain the equality $Q(s, a^*, \Psi, \Lambda, \Phi) = V(s, \Psi, \Lambda)$, meaning that now the two streams are uniquely identifiable. An alternative way that improves the stability of the training algorithm is obtained by replacing the max operation with the average advantage:

$$Q(s, a, \Psi, \Lambda, \Phi) = V(s, \Psi, \Lambda) +$$
$$+ \left( A(s, a, \Psi, \Phi) - \frac{1}{|\mathscr{A}|} \sum_{a' \in \mathscr{A}} A(s, a', \Psi, \Phi) \right). \quad (4.13)$$

In this way, when the error backpropagation is performed to update the weights, the advantages only need to change as fast as the mean instead of having to compensate for any change to the optimal action's advantage.

## 4.6 Comparison of the different networks

In the previous sections, the main architectures used in the context of DQL have been illustrated. For each of these architectures, the motivation behind them and the challenges they are supposed to address have been illustrated. For the sake of having a clear vision, this section will summarise and compare all the key features of each architecture and the most popular use cases in which they are adopted.

In chronological order, DQN stands out as one of the pioneering algorithms that introduced DNNs to RL. Its groundbreaking approach involved merging the established tabular Q-learning, traditionally employed for solving RL problems, with DNNs. This integration was specifically designed to address the challenges posed by high-dimensional state spaces. The training of such algorithm is performed through the definition of a target network, which basically is a copy of the same NN but with different coefficients updated over time. In this way, it is possible to stabilise the training process by reducing the divergence between the target $Q$-values and the predicted $Q$-values. In addition, the usage of an experience replay buffer helps break the temporal correlation between experience examples and makes the DQN capable of providing good levels of generalisation of the optimal policy. However, DQN results are prone to overestimating $Q$-values, especially in environments with complex dynamics, which may lead to sub-optimal policies.

To address the overestimation issue of DQN, the concept of double DQN was subsequently introduced. This is based on a different approach according to which the operation of action selection is decoupled from the value estimation. In particular, the main $Q$-function is used for action selection, while the target network is used for value estimation. In this way, double DQN effectively reduces the overestimation bias problem of conventional DQN, leading to more accurate value estimates.

| Algorithm | Key aspects | Weakness | Use cases |
|---|---|---|---|
| Deep Q-learning | • Possibility to handle complex, high-dimensional state spaces.<br>• Reduced temporal correlations in data by employing experience replay.<br>• Improved stability during training | • Prone to Q-values overestimation bias. | • Robotic control tasks<br>• Navigation in complex environments |
| Double deep Q-learning | • Mitigates Q-value overestimation bias by decoupling the selection and evaluation of actions | • Increased complexity | • Autonomous vehicles<br>• Real-time decision-making systems |
| Dueling deep Q-learning | • Efficient value function estimation by explicitly separating the estimation of the state value and the advantages of each action.<br>• Improved generalization | • Increased complexity | • Financial modelling<br>• Strategic planning |

*Figure 4.3 Comparative table of different DQL algorithms*

Finally, a more innovative, efficient way for value estimation and action selection has been introduced through the concept of dueling DQN. Indeed, this is a different type of approach where the $Q$-value estimation is explicitly decomposed into state value and the advantage values of each action. This allows the agent to learn which states are valuable and which actions are advantageous independently, leading to a faster convergence speed. In addition, it has also proven that the separation of value and advantage streams enhances the ability of the resulting model to generalise across different states and actions.

In summary, each of these algorithms addresses specific challenges in reinforcement learning. DQN laid the foundation of deep RL, while double DQN tackled $Q$-value overestimation, and dueling DQN introduced a more efficient way to estimate state values and action advantages. The choice among them depends on the characteristics of the environment and the specific challenges one aims to overcome. A comparative table of these algorithms is provided through Figure 4.3.

## 4.7 Summary

This chapter illustrated the most efficient and widely value function-based deep RL algorithms used nowadays. It highlighted how traditional tabular-based algorithms for RL can fall short when the dimensionality of all possible state–action pairs in the considered environment rapidly increases. To this end, deep learning has been introduced, where the tables are replaced with deep NN aimed at estimating the value function of interest. At this point, the three major algorithms currently used in the majority of work present in the literature are DQN, double DQN, and dueling DQN. For each of them, the main architecture and training mechanisms have been illustrated, as well as their main key aspect and introduced improvements. The chapter concludes with a comparative discussion of these algorithms.

# References

[1] van Hasselt H, Guez A, and Silver D. Deep reinforcement learning with double Q-learning. *AAAI Conf Artif Intell*. 2016.
[2] Wang Z, Schaul T, Hessel M, *et al. Dueling Network Architectures for Deep Reinforcement Learning*; 2015. Available from: https://arxiv.org/abs/1511.06581.

*Chapter 5*
# Policy search methods for reinforcement learning

This chapter illustrates a different type of RL technique that directly optimises the agent's policy in contrast with traditional value-based methodologies. The distinctive feature of policy search methods is that they aim to directly optimise a performance objective by finding a good policy typically represented by a DNN. Then, they are able to gracefully navigate the complexities of high-dimensional state and action spaces thanks to variants of stochastic gradient ascent with respect to the policy parameters. This chapter will illustrate and highlight the critical distinctions of the most popular policy search methods.

## 5.1 The rationale behind policy search

Within the area of RL, there has been a translation from the tabular methods initially used to find the values of each state-action pair to the use of DNN to approximate a value function and determine a policy from it. Although these value-function-based approaches have worked well in most cases, several limitations have remained in recent decades. First, due to its inner approach to solving problems, a value-based method will produce a deterministic policy. In most cases, policies are stochastic, meaning that each state's actions must have a certain probability of being selected. Furthermore, it is easy to notice how a small change in the estimated value of an action can cause the action to be selected or not, leading to convergence problems sometimes.

To solve the aforementioned issues, an alternative method was proposed in [1]. In particular, instead of approximating the value function and obtaining the optimal policy, the authors proposed to directly approximate a stochastic policy for the considered problem using an independent function approximator with its own parameters. In this case, the concept of using a DNN can be still applied, meaning that the policy can be represented through a neural network taking the state as input and providing the action probabilities as output, and the weights representing the policy parameters. To this end, we need methods for adjusting the weight of the network representing the policy so that this will converge to the optimal policy.

## 5.2 Policy gradient

The main idea of policy gradient methods is to model the policy through a DNN and optimise it by finding the optimal set of parameters $\theta$ representing the coefficients of the DNN used as the approximation function of the policy. In this context, a reward function is also defined to adjust the parameters $\theta$ subsequently. More specifically, considering an agent interacting with the environment with a policy $\pi(s, a, \theta) = Pr\{a_t = a | s_t = s, \theta\}$ for each time-step $t$, the reward function is defined as follows:

$$J(\theta) = \sum_s d^\pi(s) V^\pi(s) = \sum_s d^\pi(s) \sum_a \pi(s, a, \theta) Q^\pi(s, a), \tag{5.1}$$

where $d^{\pi_\theta}(s) = \lim_{t \to \infty} Pr\{s_t = s | s_0, \pi\}$ is the stationary distribution of states, i.e. the probability that $s_t = s$ when starting from $s_0$ and following policy $\pi_\theta$ for $t$ steps. This is supposed to exist as well as to be independent of the initial state $s_0$ for all policies.

Such reward function can be used to adjust the weights $\theta$ by using one of the gradient descent methods illustrated in the previous chapters. Then, it is necessary to develop methods to calculate $\nabla_\theta J(\theta)$, which is tricky since it depends on both the action selection and stationary state distribution. To this end, the **policy gradient theorem** comes to help. Accordingly, for any MDP, it is possible to calculate the gradient of the reward function (5.1) as

$$\nabla_\theta J(\theta) = \sum_s d^\pi(s) \sum_a Q^\pi(s, a) \nabla_\theta \pi(s, a, \theta). \tag{5.2}$$

Further details about the proof and applicability of this theorem can be found in [1]. In addition to its form in (5.2), another common representation of the gradient theorem with a slight abuse of notation is the following:

$$\begin{aligned} \nabla_\theta J(\theta) &= \sum_s d^\pi(s) \sum_a Q^\pi(s, a) \nabla_\theta \pi(s, a, \theta) \\ &= \sum_s d^\pi(s) \sum_a \pi(s, a, \theta) Q^\pi(s, a) \frac{\nabla_\theta \pi(s, a, \theta)}{\pi(s, a, \theta)} \\ &= \mathbb{E}_\pi \left[ Q^\pi(s, a) \nabla_\theta \ln(\pi(s, a, \theta)) \right]. \end{aligned} \tag{5.3}$$

It is important to point out that the expression in (5.3) is the representation of the gradient of the reward function expressed in (5.1). The majority of policy gradient methods aim at repeatedly estimating such gradients.

## 5.3 REINFORCE

The REINFORCE algorithm represents one of the fundamental methods within the family of policy gradient techniques. More specifically, it aims at estimating the gradient expressed in (5.3) through Monte Carlo simulation of different trajectories and episodes. Through this approach, an episode of a certain duration is generated by using the policy $\pi(s, a, \theta)$ and recorded as a tuple $\langle s, a, r, s' \rangle$. Once the trajectory is completely calculated, indicating with $T$ the total steps of the episode, the policy gradient is calculated as follows:

$$\nabla_\theta J(\theta) \approx \sum_{t=0}^{T} G_t \nabla_\theta \ln(\pi(s, a, \theta)), \qquad (5.4)$$

in which $G_t$ represents the discounted future rewards and is defined as $G_t = \sum_{k=0}^{T-t} \gamma^k R_{t+k}$. At this point, the gradient of the logarithmic policy is handled by specific software that implements backpropagation. Once the quantity in (5.4) is calculated, the parameters $\theta$ are updated as follows:

$$\theta \leftarrow \theta + \alpha \nabla_\theta J(\theta). \qquad (5.5)$$

Once the algorithm converges to the optimal policy, the final result will be a DNN that, for each possible state as input, will provide the probabilities of each action, assigning the highest probability to the action that will maximise the reward.

A variation of the REINFORCE algorithm, widely used to reduce the bias of the algorithm, consists of subtracting a learned baseline from the returns. Indeed, such an approach is very useful in reducing the variance of the algorithm without introducing bias in the process. Typically this baseline represents an estimate of the expected return for given a state, which can be learned alongside the policy. In this case, (5.4) assumes the form:

$$\nabla_\theta J(\theta) \approx \sum_{t=0}^{T} (G_t - b_t) \nabla_\theta \ln(\pi(s, a, \theta)), \qquad (5.6)$$

where $b_t$ represents the baseline. Alternatively, as already explained in the dueling DQL (see Chapter 4), replacing $G_t$ with the advantage function $A(s, a) = Q(s, a) - V(s)$ in (5.4) represents another way of reducing variance and limit the bias.

## 5.4 Natural policy gradient methods

The REINFORCE algorithm illustrated in the previous section is categorised as a traditional policy gradient method, which uses the gradient $\nabla_\theta J(\theta)$ to obtain the direction through which the parameters $\theta$ of the stochastic policy function $\pi(s, a, \theta)$ needs to be updated. Once the new policy is obtained a new gradient is calculated and a new adjustment is performed. The only factor that controls the size of this adjustment is the $\alpha$ factor. However, from this type of approach two main problems may arise:

- **Undershooting**: This causes small steps in the gradient direction, which can lead to slow convergence.
- **Overshooting**: Big steps may miss the reward peak and land in a sub-optimal policy region.

These problems are also common in the context of supervised learning. In that case, if we overshoot, we can correct the error later since the data has been fixed. On the other hand, in RL, data is not fixed, meaning that an overshoot may lead to a poor policy for which future sample batches may not provide much meaningful information. This can be clearly solved with a low learning rate, but it will cause slow convergence time.

To address this problem, one can think of putting a limit on the variation of the update of the policy parameter, i.e. setting the following constraint to the policy update problem:

$$\Delta\theta^* = \underset{\|\Delta\theta\|<\varepsilon}{\mathrm{argmax}}\, J(\theta + \Delta\theta). \tag{5.7}$$

This will then limit the maximum difference of updates. However, even if it sounds reasonable, this has been proven to not be enough to guarantee the reach of a suboptimal policy. To put further constraints, it is necessary to also add second-order derivatives in the updating process, which is exactly what **natural policy gradients** methods do.

In the context of natural policy gradient methods, the main focus is not to directly limit the variation of the policy parameters but to focus on the difference between the resulting policies, which implicitly act on the parameters. In literature, there are different methods for calculating the difference between probability distributions. One of these is the **Kullback–Leibner** (KL) divergence, also known as *relative entropy*, defined as follows:

$$\mathscr{D}_{\mathrm{KL}}(\pi_\theta \| \pi_\theta + \Delta_\theta) = \sum_{x \in \mathscr{X}} \pi_\theta(x) \log_2 \left( \frac{\pi_\theta(x)}{\pi_{\theta+\Delta\theta}(x)} \right), \tag{5.8}$$

This is linked with the Fisher information matrix $F(\theta)$ which describes the sensitivity of the distribution to marginal parameter changes. Suppose $\Delta\theta = 0$, we have

$$F(\theta) = \nabla_\theta^2 \mathscr{D}_{\mathrm{KL}}(\pi_\theta \| \pi_\theta + \Delta_\theta). \tag{5.9}$$

It is worth noting that if $F(\theta)$ is the identity matrix, natural and traditional policy gradient methods are equivalent. However, this is something very rare.

We can now impose the condition of the variability of policy according to the KL divergence, as follows:

$$\Delta\theta^* = \underset{\mathscr{D}_{\mathrm{KL}}(\pi_\theta \| \pi_\theta + \Delta_\theta) \leq \varepsilon}{\mathrm{argmax}}\, J(\theta + \Delta\theta). \tag{5.10}$$

In this way, it is ensured that a large update in the parameter space can be performed while guaranteeing the policy itself does not change too much. However, there is a problem related to the explicit calculation of the KL divergence since the probabilities of all action–state pairs need to be evaluated.

To simplify the calculations of the KL divergence, a method based on Lagrangian relaxation and Taylor expansion has been proposed. In particular, as illustrated in [2], (5.10) can be approximated as follows:

$$\Delta\theta^* \approx \underset{\Delta\theta}{\mathrm{argmax}}\, \nabla J(\theta)|_{\theta=\theta_{\mathrm{old}}} \cdot \Delta\theta - \frac{1}{2}\lambda \left( \Delta\theta^T F(\theta_{\mathrm{old}}) \Delta\theta \right). \tag{5.11}$$

Solving such an optimisation problem will lead to the following form of the update amount:

$$\Delta \theta = \sqrt{\frac{2\varepsilon}{\nabla_\theta J(\theta)^T F^{-1}(\theta) \nabla_\theta J(\theta)}} F^{-1}(\theta) \nabla_\theta J(\theta). \tag{5.12}$$

This expression provides the following advantages:

- It is easier to compute since the Fisher matrix can be computed as the outer product of the gradients $F(\theta) = \mathbb{E}_\theta \left[ \nabla_\theta \ln(\pi_\theta(x)) \nabla_\theta \ln(\pi_\theta(x)^T) \right]$;
- The gradient is 'corrected' by the inverse Fisher matrix, i.e. sensitivity of the policy to local changes is taken into account. More specifically, updates tend to be cautious at steep slopes and larger at flat surfaces;
- The update step $\alpha$ has a dynamic expression that adapts to local sensitivity, ensuring a policy change of magnitude $\varepsilon$ independently from the parameters.

Although these methods can ensure policy updates are stable and do not drift too far, they have numerical challenges, particularly when dealing with large-scale optimisations. However, some methods rooted in the same mathematical foundation, such as trust region policy optimisation (TRPO) and especially proximal policy optimisation (PPO), have gained much popularity.

## 5.4.1 Trust region policy optimisation

The trust region policy optimisation (TRPO) method was introduced by Schulman et al. [3] as an extension to the method proposed by Kakade & Langford [4] called conservative policy iteration, according to which the new policy was defined by the following mixture:

$$\pi_{\text{new}}(a|s) = (1 - \alpha) \pi_{\text{old}}(a|s) + \alpha \pi'(a|s), \tag{5.13}$$

where $\pi' = \arg\max_{\tilde{\pi}} L_{\pi_{\text{old}}}(\tilde{\pi})$ with

$$L_{\pi_{\text{old}}}(\tilde{\pi}) = \eta(\pi_{\text{old}}) + \sum_s \rho_{\pi_{\text{old}}}(s) \sum_a \tilde{\pi}(a|s) A_{\pi_{\text{old}}}(a|s), \tag{5.14}$$

which represents the expected return of another policy $\tilde{\pi}$ in terms of the advantage over $\pi_{\text{old}}$. In (5.14), two other quantities have been defined:

- $\eta(\pi)$ represents the expected discounted reward from the generated trajectory;
- $\rho_\pi(s) = \sum_{k=0}^{\infty} \gamma^k P(s_k = s)$ represents the discounted visitation frequencies.

Based on (5.13), the main idea of TRPO is to replace $\alpha$ with a distance measure between the *old* and *updated* policy. The measure considered is the variation divergence defined for discrete probability distributions as $D_{\text{TV}}(p||q) = \frac{1}{2} \sum_i |p_i - q_i|$. From such measure, it is possible to define:

$$D_{\text{TV}}^{\max}(\pi_{\text{old}}, \pi_{\text{new}}) = \max_s D_{\text{TV}}(\pi_{\text{old}}(\cdot|s) || \pi_{\text{new}}(\cdot|s)). \tag{5.15}$$

At this point, it has been proven that by setting $\alpha = D_{TV}^{max}(\pi_{old}, \pi_{new})$ the following bound holds:

$$\eta(\pi_{new}) \geq L_{\pi_{old}}(\pi_{new}) - \frac{4\varepsilon\gamma}{(1-\gamma)^2}\alpha^2, \tag{5.16}$$

with $\varepsilon = \max_{s,a}|A_\pi(s,a)|$. Noticing the relationship with the KL divergence, i.e. $D_{TV}(p||q)^2 \leq D_{KL}(p||q)$, (5.16) can be rewritten as follows:

$$\eta(\pi_{new}) \geq L_{\pi_{old}}(\pi_{new}) - CD_{KL}^{max}(\pi_{old}, \pi_{new}), \tag{5.17}$$

leading to the optimal value of $\theta$ increment from the maximisation problem:

$$\max_\theta [L_{\theta_{old}}(\theta) - CD_{KL}^{max}(\theta_{old}, \theta)]. \tag{5.18}$$

Although the constraint $C$ theoretically guarantees optimal step size, its usage would lead to a very small step size and a long convergence time. Practically, this is replaced with a trust region constraint on the KL divergence, leading to the following optimisation problem:

$$\begin{aligned} \max_\theta \quad & L_{\theta_{old}}(\theta) \\ \text{s.t.} \quad & CD_{KL}^{max}(\theta_{old}, \theta) \leq \delta. \end{aligned} \tag{5.19}$$

As shown in this optimisation problem, the theoretical TRPO update is not the easiest to work with. To overcome this problem, some approximations will make it more practically feasible and get an answer quickly. In particular, through some Taylor series approximation, for each episode generated, the TRPO algorithm updates the coefficients by solving the following more practical optimisation problem:

$$\begin{aligned} \theta_{k+1} = \max_\theta \quad & g(\theta_k)^T(\theta - \theta_k) \\ \text{s.t.} \quad & \frac{1}{2}(\theta - \theta_k)^T F(\theta_k)(\theta - \theta_k) \leq \delta, \end{aligned} \tag{5.20}$$

with

$$\begin{aligned} g(\theta_k) &= \frac{1}{T}\sum_t^T \nabla_\theta \ln(\pi_{\theta_k}(a_t|s_t))A_t, \\ F(\theta_k) &= \frac{1}{T}\sum_t^T \ln(\pi_{\theta_k}(a_t|s_t)) \cdot \ln(\pi_{\theta_k}(a_t|s_t))^T. \end{aligned} \tag{5.21}$$

Such a problem can be analytically solved by the methods of Lagrangian duality, yielding a solution similar to (5.12):

$$\Delta\theta = \sqrt{\frac{2\delta}{g(\theta_k)^T F^{-1}(\theta_k)g(\theta_k)}} F^{-1}(\theta_k)g(\theta_k). \tag{5.22}$$

### 5.4.2 Proximal policy optimisation

Proximal policy optimisation (PPO) is another policy gradient method recognised for its simplicity, stability, and practical implementation. Developed by OpenAI [5], it has some of the benefits of TRPO, but is simpler to implement, more general, and has better sample complexity. The main idea behind PPO is the introduction of a novel objective function with clipped probability ratios, representing a lower bound of the performance of the policy.

As illustrated in [5], before the form expressed in (5.21), the optimisation problem for the TRPO is expressed through a surrogate function is:

$$\max_{\theta} \; \mathbb{E}_t \left[ \frac{\pi_\theta(a|s)}{\pi_{\theta_{\text{old}}}(a|s)} A_t \right] \quad (5.23)$$
$$\text{s.t.} \; \mathbb{E}_t \left[ D_{\text{KL}}(\pi_{\theta_{\text{old}}}(\cdot|s), \pi_\theta(\cdot|s)) \right] \leq \delta,$$

where $A_t$ represents an estimation of the advantage function.

Indicating with $r_t(\theta) = \frac{\pi_\theta(a|s)}{\pi_{\theta_{\text{old}}}(a|s)}$, the surrogate objective function for the TRPO assumes the form of $L^{CPI}(\theta) = \mathbb{E}_t [r_t(\theta) A_t]$, which without an optimisation constraint would lead to an excessively large step. To avoid the optimisation constraint and make the algorithm more simple, the PPO introduces a clipped surrogate function:

$$L^{CLIP}(\theta) = \mathbb{E}_t \left[ \min \left( r_t(\theta) A_t, \text{clip}(r_t(\theta), 1 - \varepsilon, 1 + \varepsilon) A_t \right) \right]. \quad (5.24)$$

The term $\text{clip}(r_t(\theta), 1 - \varepsilon, 1 + \varepsilon) A_t)$ modifies the surrogate objective by clipping the probability ratio, i.e. removing the incentive for $r_t(\theta)$ moving outside of the interval $[1 - \varepsilon, 1 + \varepsilon]$.

An alternative to the clipped surrogate function would be to introduce a penalty in the function but adjust the penalty coefficient accordingly. In that case, the function to maximise will be the following:

$$L^{KLPEN}(\theta) = \mathbb{E}_t \left[ \frac{\pi_\theta(a|s)}{\pi_{\theta_{\text{old}}}(a|s)} A_t - \beta \text{KL}[\pi_{\theta_{\text{old}}}(\cdot|s_t), \pi_\theta(\cdot|s_t)] \right]. \quad (5.25)$$

In this case, the distance $d = \text{KL}[\pi_{\theta_{\text{old}}}(\cdot|s_t), \pi_\theta(\cdot|s_t)]$ is computed and the value of $\beta$ adjusted in line with a fixed target value $d_{targ}$. More specifically:

$$\begin{cases} \beta \leftarrow \beta/2 & \text{if } d < d_{targ}/1.5, \\ \beta \leftarrow 2 \times \beta & \text{if } d > 1.5 \times d_{targ}. \end{cases} \quad (5.26)$$

In both cases, $\varepsilon$ and $d_{targ}$ represent hyperparameters of the optimisation problem. In terms of implementation, in addition to the neural network which defines the policy, it involves an estimator of the advantage $A_t$. A common approach consists of running the estimated policy for $T$ time steps and collecting all the experience samples. The advantage function is estimated as follows:

$$A_t = -V(s_t) + r_t + \gamma r_{t+1} + \gamma^2 r_{t+2} + \cdots + \gamma^{T-t} V(s_T), \quad (5.27)$$

which can be generalised as (they are the same when $\lambda = 1$):

$$A_t = \delta_t + (\gamma\lambda)\delta_{t+1} + (\gamma\lambda)^2 \delta_{t+2} + \cdots + (\gamma\lambda)^{T-t+1} \delta_{T-1}, \quad (5.28)$$

where $\delta_t = r_t \gamma V(s_{t+1}) - V(s_t)$.

A very popular way for implementing PPO is through an actor-critic approach (see Algorithm 5.1). It consists of collecting $N$ different trajectories of $T$ time steps

**Algorithm 5.1** PPO implementation

**Require:** Random initialisation of DNN weights
1: **for** iteration= 1,2,··· **do**
2:     **for** agent=1,2,···,N **do**
3:         Run policy $\pi_{\theta_{\text{old}}}$ for $T$ time steps.
4:         Compute the advantage estimates $A_1,\cdots,A_T$
5:     **end for**
6:     Optimise the surrogate function with respect to $\theta$
7:     $\theta \leftarrow \theta_{\text{old}}$.
8: **end for**
9: **return** The optimal obtained policy

of data. The total $NT$ points are used to construct the surrogate function subsequently optimised through a mini-batch for $k$ epochs.

## 5.5 Deterministic policy gradient

The policy gradient methods illustrated so far aim at providing a stochastic policy, i.e. for each state, each action will have a certain probability to be performed. However, some policy gradient methods have also been developed with the aim of obtaining deterministic policies. Such policies will then provide the best action to be performed in the environment for each state observed. These are referred to as deterministic policy gradients (DPGs), where the corresponding algorithm focuses on optimising a deterministic policy to maximise the expected return.

As has already happened for the stochastic policy gradient methods, there is also a theorem that helps to make the algorithm procedure more affordable for the DPG. Indeed, starting from the state–action value function $Q(s,a)$, the DPG algorithms aim at updating the parameters of the policy function in the direction of the gradient of $Q(s,a)$. In particular, indicating with $\pi_k$ the policy at the $k$-th iteration of the algorithm, the parameters of the policy function for the $k+1$-th iteration will be updated as follows:

$$\theta^{k+1} \leftarrow \theta^k + \alpha \mathbb{E}\left[\nabla_\theta Q^{\pi_k}(s,\pi_k(s))\right]. \tag{5.29}$$

As one can notice from (5.29), the average operation $\mathbb{E}[\,\cdot\,]$ must be interpreted as an average according to the distribution probability of the state. By applying a chain rule for the calculation of the divergence, the final form for the policy gradient update rule is expressed as

$$\theta^{k+1} \leftarrow \theta^k + \alpha \mathbb{E}\left[\nabla_\theta \pi_k(s)\nabla_a Q^{\pi_k}(s,a)|_{a=\pi_k(s)}\right]. \tag{5.30}$$

Equation (5.30) seems complicated since changing the policy will potentially lead to a change of the state distribution. It is not obvious that such an approach will lead to the optimal policy. However, such result provided in (5.30) is a result form the deterministic policy gradient theorem [6]. The pseudocode for the DPG mechanism is provided in Algorithm 5.2. Regarding the $Q$-value function, this can be either fixed

**Algorithm 5.2** DPG implementation

**Require:** Random initialisation of DNN weights
1: **for** iteration $k = 1, 2, \cdots$ **do**
2:     **for** agent $= 1, 2, \cdots, N$ **do**
3:         Run policy $\pi_{\theta_k}$ for $T$ time steps.
4:         Compute the gradient as $\mathbb{E}\left[\nabla_\theta \pi_k(s) \nabla_a Q^{\pi_k}(s,a)|_{a=\pi_k(s)}\right]$
5:     **end for**
6:     Update the parameters as in (5.30).
7: **end for**
8: **return** The optimal obtained policy

or learned online with the policy. However, additional methods for executing a DPG will be illustrated in Chapter 6, in which the actor-critic approaches will be discussed.

## 5.6 Summary

This chapter has provided a brief introduction to the policy search methods, a class of RL algorithms that aim at directly finding the optimal policy instead of learning it as a greedy approach from either $V$-value or $Q$-value function. To this end, the main approaches and fundamental theorems have been exposed and illustrated, providing the reader with an essential understanding of this RL field.

## References

[1] Sutton RS, McAllester D, Singh S, *et al.* Policy gradient methods for reinforcement learning with function approximation. In: *Adv. Neural Inf. Process. Syst.*; 2000. p. 1057–1063.

[2] van Heeswijk WJA. *Natural Policy Gradients in Reinforcement Learning Explained*; 2022.

[3] Schulman J, Levine S, Moritz P, *et al. Trust Region Policy Optimization*. CoRR. 2015;abs/1502.05477.

[4] Kakade S. A natural policy gradient. In: *Adv. Neural Inf. Process. Syst.*; 2002. p. 1531–1538.

[5] Schulman J, Wolski F, Dhariwal P, *et al. Proximal Policy Optimization Algorithms*; 2017.

[6] Silver D, Lever G, Heess NMO, *et al.* Deterministic policy gradient algorithms. In: *International Conference on Machine Learning*; 2014. Available from: https://api.semanticscholar.org/CorpusID:13928442.

*Chapter 6*
# Actor-critic learning

Within the last three chapters, the reader has been introduced to value-based (Chapters 3 and 4) and policy-based (Chapter 5) methods in RL. As already anticipated in Chapter 1, in some cases, it is possible to combine the advantages from each class, leading to the class of actor-critic methods. Indeed, this architecture of RL methods consists of the actor being a policy network that outputs a probability distribution over actions and the critic in a value-based network to estimate each state-action pair's expected return. Both entities are trained jointly during the training phase, and the critic's output is used as a baseline for the actor. This chapter will provide the reader with essential knowledge on this class of RL methods that reduce the variance of policy gradient methods and improve the convergence speed.

## 6.1 Brief introduction

In the context of RL, the actor-critic architecture is gaining constant popularity. This class of RL architecture methods consists of an actor aimed at learning the policy that maximises the expected reward. By contrast, the critic will learn an accurate value function used to evaluate actions taken by the actor. Then, this can be viewed as an inner loop within the RL process, where at each step the agent obtains an improved version of the learned policy by combining its direct experience from the interaction with the environment and the feedback from the critic about the action taken. Typically, both actor and critic are implemented through DNN, with one representing the policy and another representing the value function. Then, the main questions are how these two networks are combined and the benefits of their interaction. To provide an answer to both questions, we need a brief recap on policy gradient methods. Indeed, as will be clearer later, most actor-critic algorithms result in a straightforward extension of the policy gradient methods.

### 6.1.1 *Recap on policy gradient: actor-critic approach*

In the context of policy gradient methods, the REINFORCE method, also called Vanilla policy gradient, aims to estimate the gradient of the loss function through Monte Carlo simulations. In particular, at the $k$-th iteration of $N$ parallel trajectories of $T$ time steps, the policy $\pi(s, a, \theta)$ is generated, and the approximation of the policy gradient is calculated as follows:

$$\nabla_\theta J(\theta) \approx \frac{1}{N}\sum_{i=1}^{N}\sum_{t=1}^{T}\nabla_\theta \ln(\pi(a_{i,t}|s_{i,t},\theta))G_{i,t}, \qquad (6.1)$$

where $G_{i,t} = \sum_{t'=t}^{T} r(s_{i,t'}, a_{i,t'})$ represents the cumulative reward for taking the action $a_{i,t}$ when in the state $s_{i,t}$, which represents an approximation of the $Q$-value function. Since the trajectories can greatly deviate from each other during the training phase, this results in high variability in the log probabilities and reward, which inevitably brings unstable learning and potential skewing of the policy distribution in a non-optimal direction. In addition to the issue of high gradient variance, another challenge with policy gradients arises when the trajectories yield a cumulative reward of 0. In this case, learning to be either a 'bad' or 'good' action, representing the essence of the policy gradient method, will not be performed. As already mentioned in Chapter 5, one way to reduce the variability of the gradient would consist of subtracting a baseline for the cumulative reward.

However, a better way of reducing the variability and pushing more towards the optimal policy will consist of indirectly estimating the $Q$-value function by using a DNN, which is basically the essence of the **actor-critic** approach. In this case, the $Q$-value function is parameterised with respect to other parameters $w$, leading to a new expression of the policy gradient:

$$\nabla_\theta J(\theta) \approx \frac{1}{N}\sum_{i=1}^{N}\sum_{t=1}^{T}\nabla_\theta \ln(\pi(a_{i,t}|s_{i,t},\theta))Q^w(s_{i,t}, a_{i,t}). \qquad (6.2)$$

But in this case, both the policy and value function are updated at each iteration as illustrated in Algorithm 1, in which, as already mentioned in Chapter 4, the loss function $L(w)$ used for updating the critic network is usually the square of the temporal difference $\delta_t$.

---

**Algorithm 1** $Q$ actor-critic implementation

---
**Require:** Random initialisation of DNN weights, i.e. $\theta$ and $w$.
 1: **for** iteration $k= 1,2,\cdots$ **do**
 2:     **for** agent$=1,2,\cdots,$N **do**
 3:         Run policy $\pi_{\theta_k}$ for $T$ time steps.
 4:         Sample the reward $r_t$ and the next state $s'$
 5:         Sample the next action $a' \sim \pi_{\theta_k}(a'|s')$
 6:         Update the policy parameters $\theta \leftarrow \theta + \alpha_\theta \nabla_\theta \ln(\pi(a|s,\theta_k))Q^w(s,a)$.
 7:         Compute the TD error as $\delta_t = r_t + \gamma Q^w(s',a') - Q^w(s,a)$.
 8:         Update the $Q$ function parameters as $w \leftarrow w + \alpha_w \nabla_w L(w)$.
 9:     **end for**
10: **end for**
11: **return** The optimal obtained policy

## 6.1.2 Reducing variance with the advantage

In order to reduce the variance of the policy, it is helpful to consider the advantage function $A(s,a) = Q(s,a) - V(s)$. This is because the state-value function $V(s)$ represents an optimal baseline function. Subtracting the $V$ value from the $Q$ value can be interpreted as a measure of how much better it is to take an action than the average. However, this does not mean that creating another neural network to estimate the $V$ value function is necessary. Indeed, in this case, it results in a more efficient way of exploiting the general relationship:

$$Q(s_t, a_t) = \mathbb{E}\left[r_{t+1} + \gamma V(s_{t+1})\right], \tag{6.3}$$

which allows us to calculate the advantage as follows:

$$A(s_t, a_t) = r_{t+1} + \gamma V(s_{t+1}) - V(s_t). \tag{6.4}$$

Then, estimating the $V$ value function is only necessary for implementing the entire algorithm, which is illustrated in Figure 6.1. This algorithm is commonly referred to as advantage **actor-critic (A2C)** which also has its **asynchronous version (A3C)** where the main difference is the number of critics working on estimating the global $V$ value function. More specifically, in contrast to A2C, the A3C algorithm has more than one critic networks. Although at first sight, it seems that A3C provides better performance in terms of exploration, it was empirically found by OpenAI that A2C produces comparable performance to A3C while being more efficient.

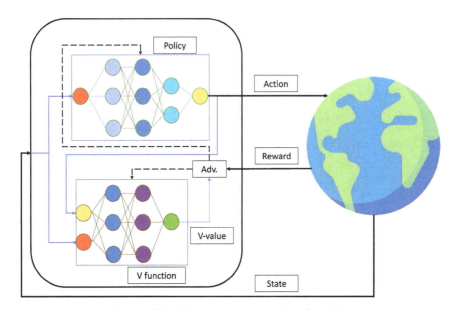

*Figure 6.1 Advantage actor-critic algorithm*

## 6.2 Different categories of methods

The previous section has outlined the main characteristics of the actor-critic approach and its benefit in improving the performance of the policy gradient algorithms. In this section, some other relevant actor-critic algorithms will be illustrated, with the main focus on those used to address real-life problems in the context of wireless communication, illustrated later in this book. A good survey of actor-critic RL can be found in [1].

### 6.2.1 Deep deterministic polity gradient

Deep deterministic policy gradient (DDPG) is a hybrid model composed of the actor part based on the value function and the critic component based on the policy search. In the DDPG algorithm, *experience replay buffer* and *target network* techniques improve convergence, speed due to reduced correlation between experience samples, and avoid excessive calculation. In the *experience replay buffer*, a finite size of memory $B$ is used to store the executed transition $\langle s^t, a^t, r^t, s^{t+1} \rangle$. After collecting enough samples, a mini-batch $D$ of transitions is randomly selected from buffer $B$ for training the neural networks. The memory $B$ is set to a finite size for updating the new sample and discarding the old ones. Instead, when the buffer is empty, the *target networks* are directly used for the critic and actor network when calculating the target value.

Let us indicate the critic network as $Q(s, a; \theta_q)$ with the parameter $\theta_q$ and the target critic network as $Q'(s, a; \theta_{q'})$ with the parameter $\theta_{q'}$. Similarly, the actor network $\mu(s; \theta_\mu)$ is initialised with the parameter $\theta_\mu$ and the target actor network $\mu'(s; \theta_{\mu'})$ with the parameter $\theta_{\mu'}$. The actor and critic network are trained using the stochastic gradient descent (SGD) over a mini-batch of $D$ samples. The critic network is updated by minimising the quantity:

$$L = \frac{1}{D} \sum_i^D \left( y^i - Q(s^i, a^i; \theta_q) \right)^2, \tag{6.5}$$

with the target

$$y^i = r^i(s^i, a^i) + \zeta Q'(s^{i+1}, a^{i+1}; \theta_{q'})|_{a^{i+1} = \mu'(s^{i+1}; \theta_{\mu'})}. \tag{6.6}$$

The actor-network parameters are updated by:

$$\nabla_{\theta_\mu} J \approx \frac{1}{D} \sum_i^D \nabla_{a^i} Q(s^i, a^i; \theta_q)|_{a^i = \mu(s^i)} \nabla_{\theta_\mu} \mu(s^i; \theta_\mu). \tag{6.7}$$

The target actor network parameters $\theta_q$ and the target critic network parameters $\theta_{\mu'}$ are updated by using soft target updates as follows:

$$\theta_{q'} \leftarrow \varkappa \theta_q + (1 - \varkappa) \theta_{q'}, \tag{6.8}$$

$$\theta_{\mu'} \leftarrow \varkappa \theta_\mu + (1 - \varkappa) \theta_{\mu'}, \tag{6.9}$$

where $\varkappa$ is a hyperparameter between 0 and 1.

In the DDPG algorithm, the deterministic policy is trained in an off-policy way; thus, for *explorations* and *exploitations* purpose, we add a noise process of $\mathcal{N}(0,1)$ as follows [2]:

$$\mu'(s^t;\theta^t_{\mu'}) = \mu(s^t;\theta^t_{\mu}) + \psi \mathcal{N}(0,1), \qquad (6.10)$$

where $\psi$ is a hyperparameter.

### 6.2.2 Asynchronous advantage actor-critic (A3C)

This algorithm represents the extension of the A2C, where the main difference consists in the fact that A3C uses multiple parallel and asynchronous threads, each of them implementing an A2C, to learn the policy. With such approach, one can easily notice how more focus in the exploration for finding the right direction for the update is provided. At the same time, the usage of multiple parallel and asynchronous threads for learning the parameters allows the removal of the *replay buffer*. A graphical representation of the A3C is provided in Figure 6.2. Typically, in either A2C or A3C, policy and value function networks are implemented through CNN with shared parameters except for the output layers. Indeed, the policy network is implemented through softmax functions while the value function is through linear function. Sometimes, an entropy factor is introduced in the policy update process, leading to the following policy update gradient:

$$\nabla_\theta \ln(\pi(a|s,\theta))A(s,a) + \beta \nabla_\theta H(\pi(s,\theta)). \qquad (6.11)$$

This term is used to penalise the model to be too sure of actions given a specific states. Indeed, the entropy represents a measure of the randomness of the policy, which will result in zero when the policy is deterministic. This term discourages the policy from collapsing into a deterministic policy too fast.

### 6.2.3 Proximal policy optimisation

By defining the policy with $\pi$ and the related parameters with $\theta_\pi$, as with the conventional PPO, here such parameters are adjusted to find an optimal policy $\pi^*$ by running the SGD over a mini-batch of $G$ transitions $(s^i, a^i, r^i, s^{i+1})$. The policy parameters are updated for optimising the objective function as follows:

$$\theta_\pi^{i+1} = \underset{\theta_\pi}{\mathrm{argmax}} \frac{1}{G}\sum_{i}^{G} \nabla_{a^i}\mathcal{L}(s^i,a^i;\theta_\pi). \qquad (6.12)$$

In the PPO algorithm, the agent interacts with the environment to find the optimal policy $\pi^*$ with the parameter $\theta_{\pi^*}$ that maximises the reward as:

$$\mathcal{L}(s,a;\theta_\pi) = \mathbb{E}\left[p_\theta^t A^\pi(s,a)\right], \qquad (6.13)$$

where $p_\theta^t = \frac{\pi(s,a;\theta_\pi)}{\pi(s,a;\theta_{old})}$ is the probability ratio of the current policy and previous policy; and $A^\pi(s,a)$ is the advantage function [3].

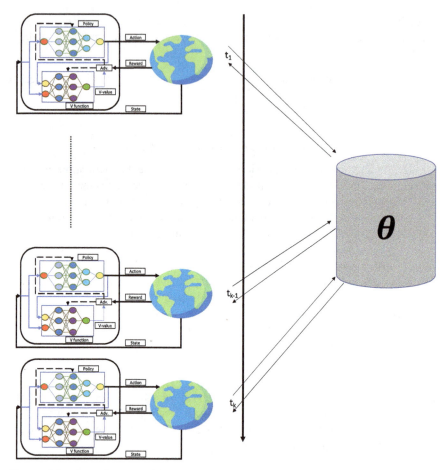

*Figure 6.2  Asynchronous Advantage Actor-Critic. In this method, the parameters $\theta$ are updated asynchronously from different parallel environments at different times.*

If we use only one network for the policy, excessive modification occurs during the training stage. Instead, the following method involving the clipping surrogate function [4] is adopted:

$$\mathscr{L}^{\text{clip}}(s, a; \theta_\pi) = \mathbb{E}\big[\min\big(p_\theta^t A^\pi(s, a), \tag{6.14}$$

$$\text{clip}\,(p_\theta^t, 1 - \varepsilon, 1 + \varepsilon) A^\pi(s, a)\big)\big], \tag{6.15}$$

where $\varepsilon$ is a small constant. In this chapter, the advantage function $A^\pi(s, a)$ [5] is formulated as follows:

$$A^\pi(s, a) = r^t + \zeta V^\pi(s^{t+1}) - V^\pi(s^t). \tag{6.16}$$

The policy is then trained by a mini-batch $B$, and the parameters are updated by:

$$\theta^{i+1} = \underset{\theta_\pi}{\arg\max}\, \mathbb{E}\big[\mathscr{L}^{\text{clip}}(s, a; \theta_\pi)\big]. \tag{6.17}$$

## 6.3 Summary

This chapter has briefly highlighted the key aspects of actor-critic-based learning methods. In particular, it started by providing a brief recap on policy gradient-based methods, underlying how these sometimes can suffer from high variability caused by the deviation of different trajectories during the training phase. To this end, it has highlighted how introducing another DNN to estimate a value function trained within the network can improve the performance of policy gradient methods. Some of the most relevant and commonly used actor-critic algorithms have also been illustrated.

## References

[1] Grondman I, Busoniu L, Lopes GAD, *et al*. A Survey of Actor-Critic Reinforcement Learning: Standard and Natural Policy Gradients. *IEEE Transactions on Systems, Man, and Cybernetics, Part C (Applications and Reviews)*. 2012;42(6):1291–1307.

[2] T. P. Lillicrap, J. J. Hunt, A. Pritzel *et al*. "Continuous control with deep reinforcement learning," in *Proc. 4th International Conf. on Learning Representations (ICLR)*, 2016.

[3] Schulman J, Moritz P, Levine S, *et al*. High-Dimensional Continuous Control Using Generalized Advantage Estimation. In: *Proc. 4th International Conf. Learning Representations (ICLR)*; 2016.

[4] Schulman J, Wolski F, Dhariwal P, *et al*. *Proximal Policy Optimization Algorithms*; 2017.

[5] V. Mnih, A. P. Badia, M. Mirza *et al.*, "Asynchronous methods for deep reinforcement learning," in *Proc. Int. Conf. Mach. Learn. PMLR*, 2016, pp. 1928–1937.

*Part III*

# Deep reinforcement learning in UAV-assisted 6G communication

*Chapter 7*
# UAV-assisted 6G communications

During the last few years, the adoption of unmanned aerial vehicles (UAVs) has emerged as keystone technology towards the successful deployment of both the fifth-generation (5G) and sixth-generation (6G) related mobile communication services. In particular, the possibility of using UAVs to extend network coverage in remote areas and provide connections during mission-critical operations has received colossal attention from the academia and industry. However, some challenges must be addressed before integrating UAVs within the next-generation networks becomes real. This chapter provides an in-depth discussion on UAV-enabled networks. In particular, it illustrates the potentialities of such network paradigm in guaranteeing error-free communication channels with sub-milliseconds communication delays, i.e. ultra-reliable low-latency communications (URLLC) necessary for the deployment of next-generation networks and services. It is concluded by pointing out some of the major open challenges that still represent a bottleneck towards the practical implementation of this promising network architecture, and providing future research directions which need to be undertaken to make UAV networks a reality*.

## 7.1 6G networks requirements

The main ingredient for developing 5G/6G networks is adopting the Internet of Things (IoT) paradigm. It is foreseen that wireless connectivity will be available for any device within the next-generation networks. This possibility will enable a wide range of new services and emerging applications such as intelligent transportation, smart manufacturing, tactile internet and virtual/augmented reality (VR/AR). The widespread use of communication devices is envisaged to foster the deployment of such innovative services, yet it comes with some implicit challenges. It is expected that such a huge diffusion of electronic devices (around $10^7$ users/km$^2$) will lead to an exponential increase of the generated data traffic up to 5016 exabytes per month [2]. Furthermore, some of these services, such as autonomous driving and VR/AR, need to experience URLLC channels, i.e. channels with end-to-end (E2E) latency less than 1 ms and a packet error probability not higher than $10^{-7}$ (see Figure 7.1) [3]. These requirements pose significant challenges. Several research efforts are underway to characterise and understand next-generation wireless communication systems.

*This chapter has been published partly in [1]

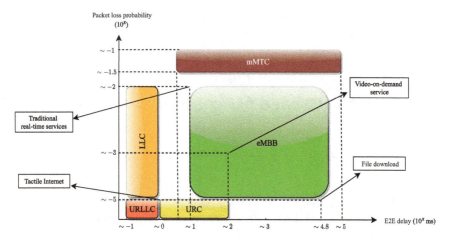

*Figure 7.1 URLLCs requirements for 6G networks*

More specifically, the research activities to guarantee improved capacity and URLLC requirements in 5G/6G networks can be classified into two main branches:

- Proposing new physical layer techniques to reduce over-the-air latency;
- Integrating innovative technologies into upper layers of the networks.

Regarding physical layer-related aspects, one way to reduce over-the-air delays is by introducing short-packet communications, which represent the typical form of traffic generated in machine-type communications and sensor networks. This is a different communication approach compared to the one adopted so far in 4G and WiFi systems, with the major implications in terms of communication theory that requires new analytical models. Under this perspective, a mathematical model for the channel capacity under the assumption of short-packet communication has been proposed in [4]:

$$R^*(k,\varepsilon) = C - \sqrt{\frac{V}{k}} Q^{-1}(\varepsilon) + \mathcal{O}\left(\frac{\log k}{k}\right), \tag{7.1}$$

in which $Q^{-1}$ denotes the inverse of the Gaussian $Q$ function, while $V = 1 - (1+\gamma)^{-2}$ and $\gamma$ denote the channel dispersion and the signal-to-noise ratio (SNR), respectively. In addition, $k$ is the packet length while $\varepsilon$ is the packet error probability. The most important aspect to notice is that the channel capacity $C = B \log_2(1+\gamma)$ is obtained from (7.1) as:

$$C = \lim_{\varepsilon \to 0} \lim_{k \to \infty} R^*(k,\varepsilon). \tag{7.2}$$

In other words, the channel capacity formula adopted so far assumes infinite packet size transmission and very low error probability, meaning that classic information-theoretic results are not applicable.

On the other hand, approaches at the upper layers are also necessary for reducing the stochastic nature of their inner delays. Indeed, most relevant operations, such as channel quality measurements, mobility management procedures, traffic prioritisation, beamforming management, handover requests, and so on, would result in huge overheads that may not comply with the low-latency requirements. For this reason, different potential methods, ranging from cross-layer design optimisation to the integration of deep learning mechanisms, have been proposed in the literature. These methods are intended to optimise the interaction between the different upper layers as much as possible to reduce the resulting overhead.

However, in addition to the need for a reduced end-to-end (E2E) delay and error probabilities, some other key aspects need to be considered in the design of URLLC-based systems.

### 7.1.1 System efficiency

With the term system efficiency, there are two levels of efficiency: (i) spectral efficiency (SE), and (ii) energy efficiency (EE). Regarding SE, this refers to the need to support certain types of services within the available bandwidth as the number of users increases, representing an urgent task. Indeed, this is a requirement for services such as autonomous driving and control of robots in smart manufacturing systems. On the other hand, other types of services based on the usage of power-constrained sensors require an increase in EE, i.e. an increase of the transmission rates with low power consumption [5,6]. Within the last few years, several optimisation algorithms have been proposed to comply with these requirements. However, they work well for small/medium-sized considered scenarios, which might not be efficient in optimising SE and EE while guaranteeing URLLC constraints when the density of users increases. Then, the main obstacle towards SE and EE achievement for such types of networks is the scalability of the optimisation problems.

### 7.1.2 Guaranteed throughput

Another requirement to consider comes from some specific services, such as tele-surgery and, in general, remote robot control, which require high levels of throughput with high availability, i.e. 99% of coverage probability during transmission. Indeed, reaching these requirements will permit high-quality maintenance and good stability of the tele-operated activities [7]. There is a huge need to improve these aspects of networks since, to date, cellular networks can achieve 95% of network availability. At the same time, existing tools are only applicable in small-scale networks [8]. This means there is a need to focus on improving the network availability by several orders of magnitude in large-scale scenarios.

## 7.2 Benefits of UAV integration

In the context of integrating innovative technologies on upper layers of the network, recently, the possibility of using UAVs to empower cellular network capabilities has been recommended by the Third Generation Partnership Project (3GPP) as a promising solution to address some 6G-related challenges [9]. Compared with conventional terrestrial base stations, in terms of on-demand deployment, UAVs are more cost-effective and more flexible. But the more important aspect of UAV-assisted communications is that they are more likely to establish short-distance line-of-sight (LoS) communication links, which in turn are able to provide better channel quality [10,11]. Due to its inner benefits, UAV-assisted cellular networks have obtained a lot of attention from the academic and industrial communities, which expect them to play an essential role in achieving high-speed wireless communication beyond 5G systems [3].

### 7.2.1 Main UAV-assisted services

So far, the vast majority of UAV-assisted applications have been focused on their usage for enhancing wireless networks' performance as a benefit of their high altitude and mobility features [1,12–31]. Some of the services which will be possible to provide through cellular-connected UAV networks are illustrated in Figure 7.2. In addition to the ones illustrated in this figure, applications of UAVs in wireless networks span across diverse research fields, such as wireless sensor networks [32], caching [33], heterogeneous cellular networks [34], massive multiple-input multiple-output (MIMO) [35], disaster communications [13,36] and device-to-device communications (D2D) [37]. For example, the UAV can be deployed to enhance the network coverage and capacity in sports events in which the existing network infrastructure cannot be boosted rapidly to meet the demand. Moreover, the UAV can provide ubiquitous wireless coverage in rural areas where terrestrial infrastructure (e.g. cables) is

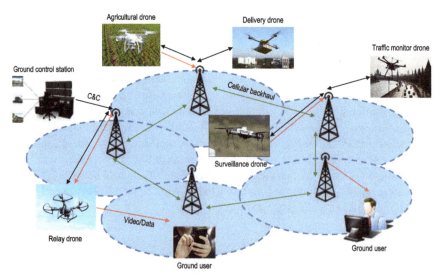

*Figure 7.2* UAV-assisted cellular network including UAV data communication link (red), cellular backhual links (green), and control and communication (C&C) links (black)

costly. In such scenarios, the UAV-enabled wireless network is an ideal solution to provide low-cost and on-demand internet to ground users.

In addition, the possibility of using UAVs for public safety communications, such as during natural disasters, has also been investigated [15,38,39]. In such critical scenarios, the terrestrial networks can be damaged and destroyed while there is a need for communication between the victims and rescue teams. Thus, the aerial network based on UAVs is a promising solution for providing a robust, fast, and on-demand communication system. The UAVs can quickly fly to the positions of ground users to provide connections. In addition, the UAVs can function as data collection machines to collect data in the IoT networks in which the ground users are limited in terms of the transmit power and communication range [1,40–45]. For example, in environments with no terrestrial infrastructure, such as remote areas and mountains, the UAVs can be deployed with emerging technologies (e.g. wireless power transfer) to provide energy-efficient and reliable communications [46–54].

Last but not least, the UAVs can be used in smart cities where deploying a base station is expensive, and the transmit signal is blocked due to high buildings and obstacles [55,56]. Clearly, UAV-assisted wireless communications can effectively establish high-speed, on-demand and cost-effective services in crowded locations or in areas poorly covered by terrestrial networks.

## 7.3 Open challenges

Although they have received a lot of attention, the vast majority of the work on UAV-enabled networks for 6G networks published in the literature has been focused on the design of optimisation frameworks aimed at mainly optimising UAV position and blocklength allocation to minimise and maximise the decoding error probability and sum data rate, respectively. However, some challenges still need to be addressed before a UAV-assisted network can be effectively deployed [27,57].

### 7.3.1 Efficient energy management

The battery capacity of UAVs plays a pivotal role in facilitating prolonged operations. As the battery capacity increases, so does the weight, leading to heightened energy consumption to complete a task. Despite significant advancements in power storage technologies that developed in recent decades, the limited availability of energy remains a substantial constraint on UAV endurance. Therefore, there is a pressing need for effective energy-conscious deployment strategies and energy-efficient operations that minimise energy consumption. One viable approach involves the collaborative efforts of multiple UAVs to enable sequential energy replenishment. For instance, wireless power transfer techniques (WPTs) can be employed for recharging drone batteries [58]. On the other hand, energy-efficient operations are focused on reducing unnecessary energy expenditure by UAVs. One way would be to perform the UAVs' movement by carefully considering the energy consumption of every manoeuvre. Another solution involves energy-efficient communications to meet the optimal trade-off that guarantees the communication requirements while minimising energy expenditure on communication-related functions such as communication circuits and signal transmission. In this scenario, a common strategy involves optimising communication approaches to maximise EE measured in bits per Joule.

### 7.3.2 3D path planning

Deploying UAVs in communication networks within a designated area presents a complex design challenge. Optimising deployments holds the potential to enhance channel gains for users, thereby improving network performance metrics. Currently, the predominant focus in research has been on single-UAV deployments within the area of interest, with only a few considering scenarios involving multiple UAVs. Compared to single-UAV scenarios, deploying multiple UAVs within the area offers superior performance, leading to improved channel quality and expanded network coverage. Indeed, cooperative deployment of multiple UAVs is a promising strategy for delivering reliable services to users. In addition, despite the capability of UAVs to dynamically adjust their three-dimensional (3D) positions swiftly for optimal performance, the current state-of-the-art (SoA) predominantly concentrates on two-dimensional (2D) placement, neglecting the more realistic 3D modelling scenarios.

Additionally, existing works primarily focus on static users, whereas ground users are in constant motion in real scenarios. In these cases, UAVs' effective path planning can significantly reduce communication distances, vital for achieving high-capacity performance. Unfortunately, determining the optimal flight path for UAVs is challenging due to the involvement of an infinite number of variables in UAV trajectory determination and practical constraints such as time-varying connectivity, fuel limitations, collision avoidance, and terrain constraints. Moreover, optimal UAV flight path is subject to a non-trivial trade-off: *increasing UAV altitude leads to higher free space path loss, but it enhances the likelihood of establishing LoS links with ground terminals.*

A common approach to UAV path planning involves approximating UAV dynamics using a discrete-time state space, where the state vector comprises position and velocity in a 3D coordinate system. The UAV trajectory, given by a sequence of states, is subject to finite transition constraints reflecting practical UAV mobility limitations. Many resulting problems with this approximation fall under the category of mixed-integer linear programming, which often proves challenging to solve.

## 7.4 Next research directions: digital twin and deep reinforcement learning

Based on the discussion provided in the previous sections, one can easily notice the necessity of intercepting the next research directions to address the current challenges that still hinder the possibility of unleashing the full potential of UAV-enabled networks. These are mainly related to the high complexity of future wireless communication networks. Indeed, due to UAV mobility, as well as the possibility of using above 6 GHz frequencies, the resulting communication channel is constantly subject to high fluctuations, which in turn require a corresponding frequent resource allocation to compensate for this fluctuation in the channel and guarantee the required levels of reliability. However, most of the current optimisation algorithms only work well for small and medium-scale problems, and sometimes, the required time does not guarantee the low-latency requirements. To this end, adopting a digital twin (DT) paradigm empowered with DRL agents has been identified as the next research direction towards achieving 6G requirements [59,60]. Indeed, the general architecture of a DT system can be represented through the three-layer hierarchical structure. More specifically:

- *Physical layer* at the bottom representing the physical world where users and services are deployed;
- *Communication layer* in the middle, which contains all the networking devices necessary for guaranteeing the data flow exchange between DT and its physical counterparts;
- *DT virtualisation and intelligence layer* on top. This layer is equipped with a storage system used to store both historical static data and dynamic data coming from the physical world, as well as different DRL agents operating on real data to find the optimal decision policy.

Thanks to this type of architecture, there is a continuous interaction between the DT, its physical counterpart, and the external environment, which allows the DT itself to remain consistently informed about real-world events. It can track the life cycle of its physical counterpart and enhance its processes and functions through closed-loop optimisation. Moreover, by employing simulations of innovative configurations and leveraging big data analytics tools, AI and DRL, the DT can forecast future states, including system defects, damages, and failures. This capability enables the proactive implementation of maintenance operations or activation of self-healing mechanisms. In the context of UAV-enabled networks, this approach, enriched with a DRL agent at the virtualisation layer, represents an innovative way to promote proactive resource optimisation while guaranteeing URLLC requirements. Indeed, the usage of DRL has already proven to be very effective in solving complex optimisation problems for UAV-assisted networks.

### 7.4.1 Resource management in UAV-enabled communications

In UAV-assisted wireless networks, DRL algorithms have shown impressive results in solving resource management problems [61–64]. In [61], cooperative UAVs were deployed to assist the cellular network. The two-step iterative algorithm using the deep Q-learning algorithm and a difference of convex algorithm was proposed to optimise the UAVs' positions, transmit beamforming and the UAV-users association to maximise the network sum rate. In [62], the UAV was deployed with the WPT technique to charge the ground devices and collect data. The multi-objective deep deterministic policy gradient method was proposed to solve the three objectives: sum-rate maximisation, harvested energy maximisation, and energy consumption minimisation.

## 7.4.2 Trajectory planning for UAV

DRL algorithms have also been applied for path planning in UAV-assisted communications [35,65–70]. In [65], the authors proposed a DRL algorithm based on the echo state network of [71] for finding the flight path, transmission power and associated cell in UAV-powered wireless networks. The deterministic policy gradient algorithm of [72] was invoked for UAV-assisted networks in [66]. The UAV's trajectory was designed to maximise the uplink sum rate without knowing the user location and the transmit power. In [35], the authors used the DQL algorithm for the UAV's navigation based on the received signal strengths estimated by a massive MIMO scheme. In [67], Q-learning was used to control the movement of multiple UAVs in the scenarios of static user locations and dynamic user locations under a random walk model. In [49], the authors characterised the DQL algorithm for minimising the data packet loss of UAV-assisted power transfer and data collection systems. The multi-agent DRL was used for trajectory design and model selection in a cellular internet of UAVs in [70]. However, the aforementioned contributions have not addressed UAV-assisted networks' joint trajectory and data collection optimisation, which is a research challenge. Furthermore, these existing works neglected interference, 3D trajectory and dynamic environment.

## 7.5 Summary

This chapter has illustrated the URLLC requirements of 6G and how these can be fulfilled by integrating UAVs into the network architecture. While UAVs provide benefits regarding channel quality and extended coverage, which can greatly foster the deployment of the innovative services included in the 6G vision, this is a research field still in its infancy stage. The main challenges and related next research directions have been outlined. Some examples of how the adoption of DRL-based approaches and their integration within a DT framework represent a promising research direction for solving the complex optimisation problems in UAV-enabled 6G networks by addressing the problems of scalability and the overall system efficiency.

## References

[1] Masaracchia A, Li Y, Nguyen KK, *et al.* UAV-Enabled Ultra-Reliable Low-Latency Communications for 6G: A Comprehensive Survey. *IEEE Access.* 2021;9:137338–137352.

[2] Saad W, Bennis M, and Chen M. A Vision of 6G Wireless Systems: Applications, Trends, Technologies, and Open Research Problems. *IEEE Netw.* 2020;34(3):134–142.

[3] Masaracchia A, Li Y, Nguyen KK, *et al.* UAV-Enabled Ultra-Reliable Low-Latency Communications for 6G: A Comprehensive Survey. *IEEE Access.* 2021;9:137338–137352.

[4] Polyanskiy Y, Poor HV, and Verdu S. Channel Coding Rate in the Finite Blocklength Regime. *IEEE Trans Inf Theory*. 2010;56(5):2307–2359.

[5] Gupta L, Jain R, and Vaszkun G. Survey of Important Issues in UAV Communication Networks. *IEEE Commun Surveys Tuts*. 2016;18(2): 1123–1152.

[6] Zhang S, Wu Q, Xu S, *et al.* Fundamental Green Tradeoffs: Progresses, Challenges, and Impacts on 5G Networks. *IEEE Commun Surveys Tuts*. 2017; 19(1):33–56.

[7] Aijaz A and Sooriyabandara M. The Tactile Internet for Industries: A Review. *Proceedings of the IEEE*. 2019;107(2):414–435.

[8] She C, Chen Z, Yang C, *et al.* Improving Network Availability of Ultra-Reliable and Low-Latency Communications With Multi-Connectivity. *IEEE Trans Commun*. 2018;66(11):5482–5496.

[9] 3GPP TSG RAN. R1-1704429 – Requirements of Connectivity Services for Drones; 2017.

[10] Mozaffari M, Saad W, Bennis M, *et al.* Mobile Unmanned Aerial Vehicles (UAVs) for Energy-Efficient Internet of Things Communications. *IEEE Trans Wireless Commun*. 2017;16(11):7574–7589.

[11] Zeng Y, Lyu J, and Zhang R. Cellular-Connected UAV: Potential, Challenges, and Promising Technologies. *IEEE Wireless Commun*. 2019;26(1): 120–127.

[12] S. Shakoor, Z. Kaleem, M. I. Baig, O. Chughtai, T. Q. Duong, and L. D. Nguyen, "Role of UAVs in public safety communications: Energy efficiency perspective," *IEEE Access*, vol. 7, pp. 140 665–140 679, Sept. 2019.

[13] Duong TQ, Nguyen LD, Tuan HD, *et al.* Learning-Aided Realtime Performance Optimisation of Cognitive UAV-Assisted Disaster Communication. In: *Proc. IEEE Global Communications Conference (GLOBECOM)*. Waikoloa, HI, USA; 2019.

[14] Nguyen LD, Nguyen KK, Kortun A, *et al.* Real-Time Deployment and Resource Allocation for Distributed UAV Systems in Disaster Relief. In: *Proc. IEEE 20th International Workshop on Signal Processing Advances in Wireless Commun. (SPAWC)*. Cannes, France; 2019. p. 1–5.

[15] Nguyen LD, Kortun A, and Duong TQ. An Introduction of Real-time Embedded Optimisation Programming for UAV Systems under Disaster Communication. *EAI Endorsed Transactions on Industrial Networks and Intelligent Systems*. 2018;5(17):1–8.

[16] Nguyen KK, Khosravirad S, Costa DBD, *et al.* Reconfigurable Intelligent Surface-assisted Multi-UAV Networks: Efficient Resource Allocation with Deep Reinforcement Learning. *IEEE J Selected Topics in Signal Process*. 2022;16(3):358–368.

[17] Zeng Y, Xu J, and Zhang R. Energy Minimization for Wireless Communication With Rotary-Wing UAV. *IEEE Trans Wireless Commun*. 2019;18(4):2329–2345.

[18] Wu Q, Zeng Y, and Zhang R. Joint Trajectory and Communication Design for Multi-UAV Enabled Wireless Networks. *IEEE Trans Wireless Commun*. 2018;17(3):2109–2121.

[19] Yang D, Wu Q, Zeng Y, *et al.* Energy Tradeoff in Ground-to-UAV Communication via Trajectory Design. *IEEE Trans Veh Technol.* 2018;67(7): 6721–6726.

[20] Zeng Y and Zhang R. Energy-Efficient UAV Communication With Trajectory Optimization. *IEEE Trans Wireless Commun.* 2017;16(6):3747–3760.

[21] Nguyen KK, Masaracchia A, Sharma V, *et al.* RIS-Assisted UAV Communications for IoT With Wireless Power Transfer Using Deep Reinforcement Learning. *IEEE J Sel Topics Signal Process.* 2022;16(5):1086–1096.

[22] Nguyen KK, Masaracchia A, Yin C, *et al.* Deep Reinforcement Learning for Intelligent Reflecting Surface-assisted D2D Communications; 2021. Available from: https://arxiv.org/abs/2108.02892.

[23] Alzenad M, El-Keyi A, Lagum F, *et al.* 3-D Placement of an Unmanned Aerial Vehicle Base Station (UAV-BS) for Energy-Efficient Maximal Coverage. *IEEE Wireless Commun Lett.* 2017;6(4):434–437.

[24] Enayati S, Saeedi H, Pishro-Nik H, *et al.* Moving Aerial Base Station Networks: A Stochastic Geometry Analysis and Design Perspective. *IEEE Trans Wireless Commun.* 2019;18(6):2977–2988.

[25] Mozaffari M, Saad W, Bennis M, *et al.* Mobile Unmanned Aerial Vehicles (UAVs) for Energy-Efficient Internet of Things Communications. *IEEE Trans Wireless Commun.* 2017;16(11):7574–7589.

[26] Jing X, Sun J, and Masouros C. Energy Aware Trajectory Optimization for Aerial Base Stations. *IEEE Trans Commun.* 2021;69(5):3352–3366.

[27] Zeng Y, Zhang R, and Lim TJ. Wireless communications with unmanned aerial vehicles: opportunities and challenges. *IEEE Commun Mag.* 2016;54(5):36–42.

[28] Wang Z, Duan L, and Zhang R. Adaptive Deployment for UAV-Aided Communication Networks. *IEEE Trans Wireless Commun.* 2019;18(9): 4531–4543.

[29] Nguyen KK, Duong TQ, Vien NA, *et al.* Non-Cooperative Energy Efficient Power Allocation Game in D2D Communication: A Multi-Agent Deep Reinforcement Learning Approach. *IEEE Access.* 2019;7:100480–100490.

[30] Nguyen KK, Vien NA, Nguyen LD, *et al.* Real-Time Energy Harvesting Aided Scheduling in UAV-Assisted D2D Networks Relying on Deep Reinforcement Learning. *IEEE Access.* 2021;9:3638–3648.

[31] Nguyen KK, Duong TQ, Vien NA, *et al.* Distributed Deep Deterministic Policy Gradient for Power Allocation Control in D2D-Based V2V Communications. *IEEE Access.* 2019;7:164533–164543.

[32] Gong J, Chang TH, Shen C, *et al.* Flight Time Minimization of UAV for Data Collection Over Wireless Sensor Networks. *IEEE J Select Areas Commun.* 2018;36(9):1942–1954.

[33] Zhong C, Gursoy MC, and Velipasalar S. Deep Reinforcement Learning-Based Edge Caching in Wireless Networks. *IEEE Trans Cogn Commun Netw.* 2020;6(1):48–61.

[34] Wu H, Wei Z, Hou Y, *et al.* Cell-Edge User Offloading via Flying UAV in Non-Uniform Heterogeneous Cellular Networks. *IEEE Trans Wireless Commun.* 2020;19(4):2411–2426.

[35] H. Huang, Y. Yang, H. Wang, Z. Ding, H. Sari, and F. Adachi, "Deep reinforcement learning for UAV navigation through massive MIMO technique," *IEEE Trans. Veh. Technol.*, vol. 69, no. 1, pp. 1117–1121, Jan. 2020.

[36] Duong TQ, Nguyen LD, and Nguyen LK. Practical Optimisation of Path Planning and Completion Time of Data Collection for UAV-enabled Disaster Communications. In: *Proc. 15th Int. Wireless Commun. Mobile Computing Conf. (IWCMC)*. Tangier, Morocco; 2019. p. 372–377.

[37] Mozaffari M, Saad W, Bennis M, *et al.* Unmanned Aerial Vehicle With Underlaid Device-to-Device Communications: Performance and Tradeoffs. *IEEE Trans Wireless Commun.* 2016;15(6):3949–3963.

[38] Wang B, Sun Y, Zhao N, *et al.* Learn to Coloring: Fast Response to Perturbation in UAV-Assisted Disaster Relief Networks. *IEEE Trans Veh Technol.* 2020;69(3):3505–3509.

[39] Zhang S and Liu J. Analysis and Optimization of Multiple Unmanned Aerial Vehicle-Assisted Communications in Post-Disaster Areas. *IEEE Trans Veh Technol.* 2018;67(12):12049–12060.

[40] Haber EE, Alameddine HA, Assi C, *et al.* UAV-Aided Ultra-Reliable Low-Latency Computation Offloading in Future IoT Networks. *IEEE Trans Commun.* 2021;69(10):6838–6851.

[41] Ranjha A and Kaddoum G. URLLC Facilitated by Mobile UAV Relay and RIS: A Joint Design of Passive Beamforming, Blocklength, and UAV Positioning. *IEEE Internet Things J.* 2021;8(6):4618–4627.

[42] Zhou F, Wu Y, Hu RQ, *et al.* Computation Rate Maximization in UAV-Enabled Wireless-Powered Mobile-Edge Computing Systems. *IEEE J Select Areas Commun.* 2018;36(9):1927–1941.

[43] Yang Z, Pan C, Wang K, *et al.* Energy Efficient Resource Allocation in UAV-Enabled Mobile Edge Computing Networks. *IEEE Trans Wireless Commun.* 2019;18(9):4576–4589.

[44] Hu Q, Cai Y, Yu G, *et al.* Joint Offloading and Trajectory Design for UAV-Enabled Mobile Edge Computing Systems. *IEEE Internet Things J.* 2019;6(2):1879–1892.

[45] Li M, Cheng N, Gao J, *et al.* Energy-Efficient UAV-Assisted Mobile Edge Computing: Resource Allocation and Trajectory Optimization. *IEEE Trans Veh Technol.* 2020;69(3):3424–3438.

[46] Yuan X, Yang T, Hu Y, *et al.* Trajectory Design for UAV-Enabled Multiuser Wireless Power Transfer With Nonlinear Energy Harvesting. *IEEE Trans Wireless Commun.* 2021;20(2):1105–1121.

[47] Hu Y, Yuan X, Zhang G, *et al.* Sustainable Wireless Sensor Networks with UAV-Enabled Wireless Power Transfer. *IEEE Trans Veh Technol.* 2021;70(8):8050–8064.

[48] Xu J, Zeng Y, and Zhang R. UAV-Enabled Wireless Power Transfer: Trajectory Design and Energy Optimization. *IEEE Trans Wireless Commun.* 2018;17(8):5092–5106.

[49] Li K, Ni W, Tovar E, *et al.* On-Board Deep Q-Network for UAV-Assisted Online Power Transfer and Data Collection. *IEEE Trans Veh Technol.* 2019;68(12):12215–12226.

[50] Yan H, Chen Y, and Yang SH. UAV-Enabled Wireless Power Transfer With Base Station Charging and UAV Power Consumption. *IEEE Trans Veh Technol*. 2020;69(11):12883–12896.

[51] Feng W, Zhao N, Ao S, *et al*. Joint 3D Trajectory Design and Time Allocation for UAV-Enabled Wireless Power Transfer Networks. *IEEE Trans Veh Technol*. 2020;69(9):9265–9278.

[52] Xie L, Xu J, and Zhang R. Throughput Maximization for UAV-Enabled Wireless Powered Communication Networks. *IEEE Internet Things J*. 2019;6(2):1690–1703.

[53] Che Y, Lai Y, Luo S, *et al*. UAV-Aided Information and Energy Transmissions for Cognitive and Sustainable 5G Networks. *IEEE Trans Wireless Commun*. 2021;20(3):1668–1683.

[54] Wu P, Xiao F, Sha C, *et al*. Trajectory Optimization for UAVs' Efficient Charging in Wireless Rechargeable Sensor Networks. *IEEE Trans Veh Technol*. 2020;69(4):4207–4220.

[55] Menouar H, Guvenc I, Akkaya K, *et al*. UAV-Enabled Intelligent Transportation Systems for the Smart City: Applications and Challenges. *IEEE Commun Mag*. 2017;55(3):22–28.

[56] Kim H, Mokdad L, and Ben-Othman J. Designing UAV Surveillance Frameworks for Smart City and Extensive Ocean with Differential Perspectives. *IEEE Commun Mag*. 2018;56(4):98–104.

[57] Mozaffari M, Saad W, Bennis M, *et al*. A Tutorial on UAVs for Wireless Networks: Applications, Challenges, and Open Problems. *IEEE Commun Surveys & Tut. 2019 Thirdquater*;21(3):2334–2360.

[58] Nguyen MT, Nguyen CV, Truong LH, *et al*. Electromagnetic Field Based WPT Technologies for UAVs: A Comprehensive Survey. *Electronics*. 2020;9(3).

[59] Masaracchia A, Sharma V, Canberk B, *et al*. Digital Twin for 6G: Taxonomy, Research Challenges, and the Road Ahead. *IEEE Open J Commun Soc*. 2022;3:2137–2150.

[60] Masaracchia A, Sharma V, Fahim M, *et al*. Digital Twin for Open RAN: Toward Intelligent and Resilient 6G Radio Access Networks. *IEEE Commun Mag*. 2023;61(11):112–118.

[61] Luong P, Gagnon F, Tran LN, *et al*. Deep Reinforcement Learning-Based Resource Allocation in Cooperative UAV-Assisted Wireless Networks. *IEEE Trans Wireless Commun*. 2021;20(11):7610–7625.

[62] Yu Y, Tang J, Huang J, *et al*. Multi-Objective Optimization for UAV-Assisted Wireless Powered IoT Networks Based on Extended DDPG Algorithm. *IEEE Trans Commun*. 2021;69(9):6361–6374.

[63] Wu F, Zhang H, Wu J, *et al*. UAV-to-Device Underlay Communications: Age of Information Minimization by Multi-Agent Deep Reinforcement Learning. *IEEE Trans Commun*. 2021;69(7):4461–4475.

[64] Ding R, Gao F, and Shen XS. 3D UAV Trajectory Design and Frequency Band Allocation for Energy-Efficient and Fair Communication: A Deep Reinforcement Learning Approach. *IEEE Trans Wireless Commun*. 2020;19(12):7796–7809.

[65] Challita U, Saad W, and Bettstetter C. Interference Management for Cellular-Connected UAVs: A Deep Reinforcement Learning Approach. *IEEE Trans Wireless Commun.* 2019;18(4):2125–2140.

[66] Yin S, Zhao S, Zhao Y, *et al.* Intelligent Trajectory Design in UAV-Aided Communications With Reinforcement Learning. *IEEE Trans Veh Technol.* 2019;68(8):8227–8231.

[67] Liu X, Liu Y, and Chen Y. Reinforcement Learning in Multiple-UAV Networks: Deployment and Movement Design. *IEEE Trans Veh Technol.* 2019;68(8):8036–8049.

[68] Wang C, Wang J, Shen Y, *et al.* Autonomous Navigation of UAVs in Large-Scale Complex Environments: A Deep Reinforcement Learning Approach. *IEEE Trans Veh Technol.* 2019;68(3):2124–2136.

[69] Samir M, Assi C, Sharafeddine S, *et al.* Age of Information Aware Trajectory Planning of UAVs in Intelligent Transportation Systems: A Deep Learning Approach. *IEEE Trans Veh Technol.* 2020;69(11):12382–12395.

[70] Wu F, Zhang H, Wu J, *et al.* Cellular UAV-to-Device Communications: Trajectory Design and Mode Selection by Multi-Agent Deep Reinforcement Learning. *IEEE Trans Commun.* 2020;68(7):4175–4189.

[71] Jaeger H. The "echo state" approach to analysing and training recurrent neural networks-with an erratum note. GMD – German National Research Institute for Computer Science, Tech Rep. 2010;148(34):13.

[72] T. P. Lillicrap, J. J. Hunt, A. Pritzel *et al.* "Continuous control with deep reinforcement learning," in *Proc. 4th International Conf. on Learning Representations (ICLR)*, 2016.

*Chapter 8*
# Distributed deep deterministic policy gradient for power allocation control in UAV-to-UAV-based communications

Nowadays, UAV-to-UAV (U2U) communication represents an emerging revolutionary technology that is envisioned to play a pivotal role in the evolution of communications beyond 5G network-enabled communications. Indeed, it represents a core paradigm for the next generation of platforms and applications, such as real-time high-quality video streaming, virtual reality games, and smart city operations. However, the rapid proliferation of users that these networks will need to serve, as well as the number of sensors that each UAV can support, represents a call for more efficient resource allocation algorithms to enhance network performance while still being capable of guaranteeing the quality of service. The adoption of DRL is emerging as a powerful tool to enable each node in the network to have a real-time self-organising ability. This chapter presents two novel approaches based on deep deterministic policy gradient algorithm, namely 'distributed deep deterministic policy gradient' and 'sharing deep deterministic policy gradient', for the multi-agent power allocation problem in U2U-based communications. Numerical results highlight how the proposed models outperform other DRL approaches regarding the network's energy efficiency and flexibility.*

## 8.1 Introduction

UAV-to-UAV (U2U) communication, which utilises intelligent autonomous drones to provide different types of task-oriented services, has recently emerged as a promising technology. Similar to device-to-device (D2D) communications, U2U networks consist of a swarm of multiple drones cooperating to achieve a particular task such as extending the network coverage in remote areas, vehicle traffic monitoring and surveillance [2].

In D2D communications, end users can interact with each other without having to connect directly to base stations (BS) or core networks. It enables the development of various platforms and applications. For example, D2D communication

---

*This chapter has been published partly in [1].

is a core technique in smart cities [3], high-quality video streaming [4], and disaster relief networks [5]. D2D communication can also support vehicle-to-vehicle (V2V) communications as it has tremendous advantages such as spectral efficiency, energy efficiency, and fairness [6–9]. First, the V2V communications under the D2D-enabled architecture are supported through localised D2D communication to inherit the benefits of D2D-based networks. Techniques used in D2D communication substantially reduce latency and power consumption; hence, they are suitable for tight delay V2V communications. Second, the time constraint requirement in V2V links is as strict as in D2D pairs due to the low latency, which is essential for critical safety services. In addition, the demand for high reliability in V2V communication is approximately similar to D2D communication. The V2V link reliability is guaranteed by ensuring the SINR is not lower than a small threshold. We identify and incorporate the reliability QoS requirements for V2V links into the objective formulation. Therefore, D2D communication is an emerging solution enabling safe, efficient, and reliable V2V communications. However, the resource allocation problem is one of the challenges to enabling D2D-based V2V communications due to rapid channel variations caused by V2V user mobility.

Resource allocation problems in D2D communication have received enormous attention from the research community [10–15]. In [10], the authors considered three scenarios, namely the perfect channel state information, partial channel state information, and the imperfect channel between the users and the transmitters, to present a resource allocation algorithm to achieve optimal performance in terms of secrecy throughput and energy efficiency. In [11], the authors introduced an optimisation scheme based on the combination of coral reefs optimisation and quantum evolution to gain optimal results for joint resource management and power allocation problems in cooperative D2D heterogeneous networks. The authors in [12] proposed an optimisation algorithm based on logarithm inequality to solve the joint energy-harvesting time and power allocation in D2D communications assisted by unmanned aerial vehicles. Meanwhile, in [13], to maximise the total average achievable rate from D2D transmitters to D2D receivers, the authors proposed an optimal solution to allocate the spectrum and power in cooperative D2D communications with multiple D2D pairs. In [14], a resource allocation approach was presented to improve energy-efficient D2D communication. The power allocation problem was solved using the Lambert W function, and channel allocation was solved appropriately by the Gale-Shapley matching algorithm. However, all the above approaches have a common drawback that requires the data of all D2D pairs to be collected and processed in a centralised manner at the BS. It causes delays in real-time scenarios. Furthermore, many previous algorithms typically only worked in a small, static environment, and all the data was analysed at one point. It is unrealistic because environments are dynamic, and centralised processing will inflict bottlenecks, congestion, and blockage at the BS or central processing unit.

Some recent works have studied to apply techniques in D2D communication to support V2V communications [6–9]. In [7], a cluster-based resource block sharing algorithm and in [9], a separate resource block algorithm was proposed to deal with the radio resource allocation problem in D2D-based vehicle-to-everything communications. Meanwhile, the authors in [8] proposed a grouping algorithm, channel selection, and power control strategies to maximise the performance of a network consisting of multiple D2D-based V2V links sharing the same channel. However, the major issue in both D2D and U2U communication is that each device in the network typically has limited resources and power for transmitting information while the demand for efficient resource allocation, such as spectrum and power allocation, is rising rapidly. Furthermore, each device in either D2D or U2U networks cannot frequently transfer or store in its memory the information of its resource allocation scheme due to limitations in transmission power and memory storage. Besides, if we use BS as a central processing unit to find a resource allocation scheme for each pair, the delay incurred will make the system model unsuitable for real-time applications. Efficient optimisation algorithms have recently been deployed to enhance energy efficiency and processing time [12,16,17].

In [18], a RL algorithm was used to obtain the optimal policy for the power control problem in energy-harvesting two-hop communication. The authors considered that each energy-harvesting node only knows the harvested energy and channel coefficients. Thus, the problem can be transferred to two point-to-point problems, and to maximise the amount of data at the receiver, a RL algorithm called SARSA is employed at each energy-harvesting node to reach the optimal policy at a transmitter. Nevertheless, the RL-based algorithm has some disadvantages, such as instability and inefficiency, when the number of nodes in the network is sufficiently large.

Recently, deep learning (DL), a subfield of machine learning, has been considered a powerful optimisation tool to solve resource management problems in modern wireless networks [19,20]. An approach based on deep recurrent neural networks was presented in [19] to obtain the optimal policy for resource allocation in a non-orthogonal multiple access-based heterogeneous internet-of-things network. In [20], the authors proposed a DL-based resource management scheme to balance cognitive radio networks' energy and spectrum efficiency. The convergence speed was significantly improved by utilising the neural networks regarding the lower computational complexity and learning cost while satisfying the network performance. DL has also been applied to solve the physical layer issues in wireless networks [21–25]. The authors in [21] proposed a convolutional neural network-based method to automatically recognise eight popular modulation models used in advanced cognitive radio networks. The proposed network was trained by using the two datasets of in-phase and quadrature to extract features and efficiently classify modulated signals. Meanwhile, the authors in [22] introduced a fully connected neural network-based framework for maximising the network throughput under the limited constraint of total transmit power. The data was generated without labels and put into the neural network for offline unsupervised training. The DL-based algorithms were also proposed to enable mmWave massive multiple-input multiple-output framework for hybrid precoding schemes [23] and to detect the channel characteristics automatically [24].

DRL, a combination of RL and deep neural networks, has been used widely in wireless communication thanks to its powerful features, impressive performance, and adequate processing time. The authors in [26] formulated a non-cooperative power allocation game in D2D communications and proposed three approaches based on deep Q-learning, double deep Q-learning, and dueling deep Q-learning algorithm for multi-agent learning to find the optimal power level for each D2D pair to maximise the network performance. The authors in [27] used a deep Q-learning algorithm to look for the optimal sub-band and transmission power level for each V2V user in V2V communications while satisfying the low latency requirement. However, these algorithms can only work on the discrete action space; hence, human intervention is required to design the power level of each pair. With the finite set of action space, the performance of these algorithms cannot reach the optimal result, and the reward can become worse if we cannot divide the power level accurately.

Against this background, in this chapter, we propose two novel models termed distributed deep deterministic policy gradient (DDDPG) and sharing deep deterministic policy gradient (SDDPG) based on deep deterministic policy gradient (DDPG) algorithm [28]. The proposed approaches can work on a continuous action space for the multi-agent power allocation problem in U2U-based communications. Therefore, we can improve the algorithm convergence quality and sample efficiency significantly, especially when the number of U2U pairs in the network increases. We will further show that the numerical results of our model outperform the approach based on the original DDPG algorithm in terms of energy efficiency (EE) performance, computational complexity, and network flexibility. Our main contributions are as follows:

- We provide two novel approaches based on a deep deterministic policy gradient algorithm to solve the multi-agent learning and non-cooperative power allocation problem in U2U-based communications. Experiment results show promising results over other existing deep reinforcement learning approaches.
- By modifying the input of the neural network, all the multi-agent deep reinforcement learning algorithm agents can share one actor network and one critic network to reach higher performance and faster convergence while reducing the computational complexity and memory storage significantly.
- Finally, after training the policy neural network, the non-cooperative power allocation problem in U2U-based communications can be solved in milliseconds. It becomes a promising technique for real-time scenarios.

The remainder of the chapter is organised as follows. In Section 8.2, we describe the system model and formulation of the multi-agent power allocation problem in U2U-based communications. Section 8.3 describes the value functions and policy gradient concepts and proposes a distributed deep deterministic policy gradient algorithm-based method. In Section 8.4, we improve the model by using the embedding layer to solve the non-cooperative resource allocation problem in D2D-based V2V communications efficiently. In Section 8.5, the simulation results are presented to demonstrate the efficiency of our proposed schemes. Finally, we conclude this chapter and propose future works in Section 8.6.

## 8.2 System model and problem formulation

This section defines the system model and formulation of the power allocation problem in U2U-based communications. As depicted in Figure 8.1, $N$ U2U pairs are

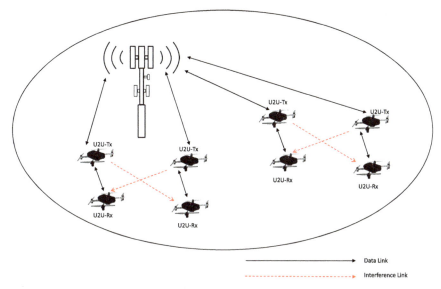

*Figure 8.1  System model of U2U-based communications*

distributed randomly within the coverage of one BS. Each U2U pair comprises a single antenna U2U transmitter (U2U-Tx) and a single antenna U2U receiver (U2U-Rx). Within this model $\beta_0, f_i$, and $\alpha_h$ represent the channel's power gain at the reference distance, an exponentially distributed random variable with unit mean, and the path loss exponent for U2U links, respectively. The location of the $i$th U2U-Tx and $j$th U2U-Rx with $i, j \in \{1, \ldots, N\}$ are $(x^i_{Tx}, y^i_{Tx})$ and $(x^j_{Rx}, y^j_{Rx})$. Hence, the channel power gain $h_{ij}$ between the $i$th U2U-Tx and $j$th U2U-Rx is expressed as:

$$h_{ij} = \beta_0 f_i^2 R_{ij}^{-\alpha_h}, \tag{8.1}$$

where $R_{ij} = \sqrt{(x^i_{Tx} - x^j_{Rx})^2 + (y^i_{Tx} - y^j_{Rx})^2}$ is the Euclidean distance between the $i$th U2U-Tx and $j$th U2U-Rx, in which we assume that UAV stays the same altitude $H_z$. The received signal-to-interference-plus-noise ratio (SINR) at the $i$th U2U user is defined as:

$$\gamma_i = \frac{p_i h_{ii}}{\sum_{j \in N}^{j \neq i} p_j h_{ji} + \sigma^2}, \tag{8.2}$$

where $p_i \in (p_i^{min}, p_i^{max})$ and $\sigma$ are the transmission power at $i$th U2U pairs and the AWGN power, respectively.

In the power allocation problem in U2U-based communications with $N$ U2U pairs, our objective is to find an optimal policy to maximise the EE performance of the considered network. The information throughput at the $i$th U2U pair is defined as:

$$\begin{aligned}\psi_i &= W \ln(1 + \gamma_i) \\ &= W \ln\left(1 + \frac{p_i h_{ii}}{\sum_{j \in N}^{j \neq i} p_j h_{ji} + \sigma^2}\right),\end{aligned} \tag{8.3}$$

where $W$ represents bandwidth. The total network performance is a joint function of all U2U pairs. We define the quality of service (QoS) constraints as:

$$\gamma_i \geq \gamma_i^*, \forall i \in N. \tag{8.4}$$

In this work, we focus on maximising the total EE performance of the network while satisfying energy constraints and the QoS constraints for each U2U pair. Therefore, the EE optimisation problem can be defined as:

$$\max \sum_i^N \frac{W}{p_i} \ln\left(1 + \frac{p_i h_{ii}}{\sum_{j \in N}^{j \neq i} p_j h_{ji} + \sigma^2}\right), \tag{8.5}$$

$$s.t \quad \gamma_i \geq \gamma_i^*, \forall i \in N, \tag{8.6}$$

$$p_i^{min} \leq p_i \leq p_i^{max}. \tag{8.7}$$

In the U2U-based communications, we have $N$ U2U pairs in which each U2U pair can only have its environment information about power allocation strategy and current environment state. This makes the power allocation problem in U2U-based communications a multi-agent and non-cooperative game. Thus, we formulate the multi-agent power allocation game in U2U-based communications and propose two deep reinforcement learning approaches based on the DDPG algorithm to enable each U2U user to have an optimal power allocation scheme.

In RL, an agent interacts with the environment to find the optimal policy through trial-and-error learning. We can formulate this task as a Markov decision process (MDP) [29]. In particular, we define a 4-tuple $\langle \mathscr{S}, \mathscr{A}, \mathscr{R}, \mathscr{P} \rangle$, where $\mathscr{S}$ and $\mathscr{A}$ are the agent state space and action space, respectively. The reward function $r = \mathscr{R}(s, a, s')$ can be obtained at state $s \in \mathscr{S}$, action $a \in \mathscr{A}$, and next state $s' \in \mathscr{S}$. An agent has transition function $\mathscr{P}_{ss'}^a$ which is the probability of next states $s'$ when taking action $a \in \mathscr{A}$ at state $s \in \mathscr{S}$.

Regarding the multi-agent power allocation problem in U2U-based communications, we define each U2U transmitter as an agent, with the entire system consists of $N$ agents. The $i$th U2U-Tx is defined as $i$th agent, which is represented as $\langle \mathscr{S}_i, \mathscr{A}_i, \mathscr{R}_i, \mathscr{P}_i \rangle$, where $\mathscr{S}_i$ is the environment state space, $\mathscr{A}_i$ is the action space, $\mathscr{R}_i$ is the reward function, and $\mathscr{P}_i$ is the state transition probability function. Generally, an agent corresponding to a U2U user at each time $t$ observes a state, $s^t$ from the state space, $\mathscr{S}$, then accordingly takes the action of selecting power level, $a^t$, from the action space, $\mathscr{A}$ based on the policy, $\pi$. By taking the action $a^t$, the agent receives a reward, $r^t$, and the environment transits to a new state $s^{t+1}$.

In the next step, we define the action spaces, state spaces and reward function of the multi-agent power allocation problem in U2U-based communications as follows:

**State spaces**: At each time $t$, the state space of the $i$th U2U transmitter observed by the U2U link for characterising the environment is defined as:

$$\mathscr{S}_i = \{i, \mathscr{I}_i\}, \tag{8.8}$$

where $\mathscr{I}_i \in (0, 1)$ is the level of interference as:

$$\mathscr{I}_i = \begin{cases} 1 & \text{for} \quad \gamma_i \geq \gamma_i^* \\ 0 & \text{for} \quad \text{otherwise} \end{cases} \tag{8.9}$$

**Action spaces**: The agent $i$ at time $t$ takes an action $a_i^t$, which represents the agent selected power level, according to the current state, $s_i^t \in \mathscr{S}_i$ under the policy $\pi_j$. The action space of $i$th U2U-Tx is denoted as:

$$\mathscr{A}_i = \{p_i\}, \tag{8.10}$$

where $p_i^{\min} \leq p_i \leq p_i^{\max}$.

**Reward function**: Our objective is to maximise the total performance of the network by interacting with the environments while satisfying the QoS constraints. Thus, we design a reward function $\mathscr{R}_i$ of the $i$th U2U user in state $s_i$ by receiving the immediate return by executing action $a_i$ as:

$$\mathscr{R}_i = \begin{cases} \frac{W}{p_i} \ln(1+\gamma_i) & \text{if } \mathscr{I}_i = 1 \\ 0 & \text{if } \mathscr{I}_i = 0 \end{cases} \tag{8.11}$$

## 8.3 Multi-agent power allocation problem in U2U-based communications: distributed deep deterministic policy gradient approach

RL has two main approaches and a hybrid model to solve the games. There are value function-based methods, policy search-based methods, and an actor–critic approach that employs both value functions and policy search [30]. This section explains value function and policy search concepts that can be learned on continuous domains. We propose a solution based on the DDPG algorithm to solve the energy-efficient power allocation problem in U2U-based communications.

### 8.3.1 Value function

Value function, which is often denoted as $V^\pi(s)$, estimates the expected reward for an agent starting in state $s$ and following the policy $\pi$ subsequently. Value function represents how good an agent is in a given state and is expressed as:

$$V^\pi(s) = \mathbb{E}\big[\mathscr{R}|s_0 = s, \pi\big], \tag{8.12}$$

where $\mathbb{E}(\cdot)$ stands for the expectation operation and $\mathscr{R}$ denotes the rewards gain from the initial state $s$ while following the policy $\pi$. In all the possibility of the value function $V^\pi(s)$ there is an optimal value $V^*(s)$ corresponding to an optimal policy $\pi^*$; the optimal value function $V^*(s)$ can be defined as:

$$V^*(s) = \max_\pi V^\pi(s), s \in \mathscr{S}. \tag{8.13}$$

The optimal policy $\pi^*$ is the policy that can be retrieved from optimal value function $V^*(s)$ by choosing the action $a$ from the given state $s$ to maximise the expected reward. We can rewrite (8.13) by using Bellman equation [31] as follows:

$$V^*(s) = V^{\pi^*}(s) = \max_{a \in \mathscr{A}} \left[ r(s,a) + \zeta \sum_{s' \in \mathscr{S}} p_{ss'}^a V^*(s') \right], \tag{8.14}$$

where $r(s,a)$ is the expected reward obtain when taking action $a$ from the state $s$, $p_{ss'}^a$ defines the probability of the next state $s'$ if the agent at the state $s$ takes action $a$, and $\zeta \in [0,1]$ is the discount factor.

The action-value function $Q^\pi(s,a)$ is the total reward representing how good it is for an agent to pick an action $a$ in state $s$ when following policy $\pi$:

$$Q^\pi(s,a) = \mathbb{E}\big[r(s,a) + \zeta \mathbb{E}[V^\pi(s')]\big]. \tag{8.15}$$

The optimal action-value function $Q^*(s,a)$ can be written as:

$$Q^*(s,a) = \mathbb{E}\Big[r(s,a) + \zeta \sum_{s' \in \mathcal{S}} p_{ss'}^a V^*(s')\Big]. \tag{8.16}$$

Thus, we have

$$V^*(s) = \max_{a \in \mathcal{A}} Q^*(s,a). \tag{8.17}$$

$$Q^*(s,a) = \mathbb{E}\big[r(s,a) + \zeta \max_{a' \in \mathcal{A}} Q^*(s',a')\big]. \tag{8.18}$$

Q-learning, an off-policy algorithm regularly uses the greedy policy $\pi = \text{argmax}_{a \in \mathcal{A}} Q(s,a)$ to choose the action. The agent can achieve the optimal results by adjusting the $Q$ value according to the updated rule:

$$Q(s,a) \leftarrow (1-\alpha)Q(s,a) + \alpha\big[r(s,a) + \zeta \max_{a' \in \mathcal{A}} Q(s',a')\big], \tag{8.19}$$

where $\alpha \in [0,1]$ is the learning rate.

### 8.3.2 Policy search

The policy gradient, which is one of the policy search techniques, is a gradient-based optimisation algorithm. It aims to model and optimise the policy to directly search for an optimal behaviour strategy $\pi^*$ for the agent. The policy gradient method is popular because of the efficient sampling ability when the number of policy parameters is large. Let $\pi$ and $\theta_\pi$ denote the policy and vector of policy parameters, respectively, and $J$ is the performance of the corresponding policy. The value of the reward function depends on this policy, and then the various algorithms can be applied to optimise parameter $\theta_\pi$ to achieve the optimal performance.

The average reward function on MDPs can be written as:

$$J(\theta) = \sum_{s \in \mathcal{S}} d(s) \sum_{a \in \mathcal{A}} \pi_\theta(s,a;\theta_\pi) \mathcal{R}(s,a), \tag{8.20}$$

where $d(s)$ is the stationary distribution of Markov chain for policy $\pi_\theta$. Using gradient ascent, we can adjust the parameter $\theta_\pi$ suggested by $\nabla_\theta J(\theta_\pi)$ to find the optimal $\theta_\pi^*$ that produces the highest reward. The policy gradient can be computed in [32] as follows:

$$\nabla_\theta J = \sum_{s \in \mathcal{S}} d(s) \sum_{a \in \mathcal{A}} \pi_\theta(s,a;\theta_\pi) \nabla_\theta \log \pi_\theta(s,a;\theta_\pi) Q^\pi(s,a)$$
$$= \mathbb{E}_{\pi_\theta}\big[\nabla_{\theta_\pi} \log \pi_\theta(s,a;\theta_\pi) Q^\pi(s,a)\big]. \tag{8.21}$$

The REINFORCE algorithm is devised as a Monte-Carlo policy gradient learning algorithm that relies on an estimated return by Monte-Carlo simulations where episode samples are used to update the policy parameter $\theta_\pi$. The objective of the REINFORCE algorithm is to maximise expected rewards under policy $\pi$:

$$\theta_\pi^* = \underset{\theta_\pi}{\mathrm{argmax}}\, J(\theta). \tag{8.22}$$

Thus, the gradient is presented as:

$$\nabla \theta_\pi = \mathbb{E}_{\pi_\theta}\left[\nabla_{\theta_\pi} \log \pi_\theta(s,a;\theta_\pi) Q^\pi(s,a)\right]. \tag{8.23}$$

Then, parameters are updated along a positive gradient direction:

$$\theta_\pi \leftarrow \theta_\pi + \alpha \nabla \theta_\pi. \tag{8.24}$$

A drawback of the REINFORCE algorithm is the slow speed of convergence due to the high variance of the policy gradients.

### 8.3.3 Distributed deep deterministic policy gradient

By utilising the advantages of both policy search-based methods and value function-based methods, a hybrid model called the actor-critic algorithm has grown as an effective approach [30]. In policy gradient-based methods, the policy function $\pi(a|s)$ is always modelled as a probability distribution over actions space $\mathscr{A}$ in the current state, and thus it is stochastic. Very recently, deterministic policy gradient (DPG) has been deployed as an actor–critic algorithm in which the policy gradient theorem is extended from stochastic policy to deterministic policy. Inspired by the success of deep Q-learning [26], which uses neural network function approximation to learn value functions for a very large state and action space online, the combination of DPG and DL called deep deterministic policy gradient enables learning in continuous spaces.

An existing drawback of most optimisation algorithms is that the samples are assumed to be independently and identically distributed. It leads to the destabilisation and divergence of RL algorithms if we use a non-linear approximate function. To overcome that challenge, we use two major techniques:

- *Experience replay buffer*: agent $i$ has a replay buffer $\mathscr{D}_i$ to store the samples and take mini-batches for training. Transitions are sampled from the environment, following the exploration policy, and the tuple $(s_i^t, a_i^t, r_i^t, s_i^{t+1})$ will be stored in $\mathscr{D}_i$. When the replay buffer $\mathscr{D}_i$ is big enough, a mini-batch $K_i$ of transitions is sampled randomly from the buffer $\mathscr{D}_i$ to train the actor and critic network. By setting the finite size of replay buffer $\mathscr{D}$, the oldest samples are removed to retrieve space for the new samples, and the buffers are always up to date.
- *Target network*: At each step of training, the $Q$ value is shifted. Thus, if we use a constantly shifting set of values to estimate the target value, the value estimations are easily out of control, and it makes the network unstable. To address this issue, we use a copy of the actor and critic networks, $Q_i'(s_i, a_i; \theta_{q_i'})$ and $\mu_i'(s_i; \theta_{\mu_i'})$, respectively, to calculate the target values. The parameters $\theta_{q_i'}$

and $\theta_{\mu'_i}$ in the actor and critic network are then updated using soft target updates with $\tau \ll 1$:

$$\theta_{q'_i} \leftarrow \tau\theta_{q_i} + (1-\tau)\theta_{q'_i} \qquad (8.25)$$

$$\theta_{\mu'_i} \leftarrow \tau\theta_{\mu_i} + (1-\tau)\theta_{\mu'_i} \qquad (8.26)$$

By using the target networks, the target values are constrained to change slowly, significantly learning the action-value function closer to supervised learning. However, both targets $\mu'_i$ and $Q'_i$ are required to process a stable target to train the critic consistently without divergence. Herein, this may slow training since the target network delays the propagation of value estimations.

A notable challenge of learning in continuous action spaces is exploration [28]. To do better exploration, we add a small white noise $\mathcal{N}_i(0,1)$ to our actor policy to construct a Gaussian exploration policy $\mu'_i$ [28]:

$$\mu'_i(s^t_i) = \mu_i(s^t_i; \theta^t_{\mu_i}) + \varepsilon \mathcal{N}_i(0,1), \qquad (8.27)$$

where $\varepsilon$ is a small positive constant. The details of our proposed algorithm, distributed deep deterministic policy gradient based on the DDPG algorithm, to deal with the multi-agent power allocation problem in D2D-based V2V communications are described in Algorithm 8.1.

## 8.4 Sharing deep deterministic policy gradient for multi-agent power allocation problem in D2D-based V2V communications

In this section, we present a simple improvement of the DDPG algorithm with the parameter-sharing technique in multi-agent learning problems. This algorithm can reach more effective policies for all the V2V pairs in the network by sharing the parameters of a single policy due to the homogeneous quality of all agents. Therein, each agent can be trained with the experiences of all agents simultaneously [33].

With the DDDPG algorithm in Algorithm 8.1, each agent has an actor network and a critic network of their own. It makes the systems shift significantly when the number of U2U pairs increases. In addition, the computational complexity, memory storage, and processing time are also unmanageable. Inspired by the impressive results of the paper [33], to overcome that problem, we propose a novel model based on DDPG called the SDDPG algorithm in which a large number of agents can use sharing networks. By adding the embedding layer to build a new input layer of neural networks, we can use one actor and one critic network for the multiple agents in deep reinforcement learning. Consequently, it significantly reduces our model's overall computational processing while ensuring performance. The speed of convergence is also better than standard approaches.

The simplest way to represent an input layer with a node for every pair is 'one-hot' encoding, which is a vector of zeros with one at a single position. However, when the number of U2U pairs in the network increases, the 'one-hot' encoding vector becomes more sparse with relatively few non-zero values. Thus, 'one-hot'

**Algorithm 8.1** Distributed deep deterministic policy gradient algorithm for multi-agent power allocation problem in U2U-based communications

1: Initialisation:
2: **for** U2U $i, i \in N$ **do**
3:     Randomly initialise critic $Q_i(s_i, a_i; \theta_{q_i})$ and actor $\mu_i(s_i; \theta_{\mu_i})$
4:     Randomly initialise targets $Q'_i$ and $\mu'_i$ with parameter $\theta_{q'_i} \leftarrow \theta_{q_i}, \theta_{\mu'_i} \leftarrow \theta_{\mu_i}$
5:     Initialise replay buffer $D_i$
6: **end for**
7: **for** V2V $i, i \in N$ **do**
8:     **for** episode $= 1, \ldots, M$ **do**
9:         Initialise the action exploration to a Gaussian $\mathcal{N}_i$
10:         Receive initial observation state $s_i^1$
11:         **for** iteration $= 1, \ldots, T$ **do**
12:             Obtain the action $a_i^t$ at state $s_i^t$ according to the current policy and action exploration noise
13:             Measure the achieved SINR at the receiver according to (8.2)
14:             Update the reward $r_i^t$ according to (8.11)
15:             Observe the new state $s_i^{t+1}$
16:             Store transition $(s_i^t, a_i^t, r_i^t, s_i^{t+1})$ into replay buffer $D_i$
17:             Sample randomly a mini-batch of $K_i$ transitions $(s_i^k, a_i^k, r_i^k, s_i^{k+1})$ from buffer $D_i$
18:             Update critic by minimising the loss:

$$L_i = \frac{1}{K_i} \sum \left( y_i^k - Q_i(s_i^k, a_i^k; \theta_{q_i}) \right)^2, \quad (8.28)$$

where

$$\begin{aligned} y_i^k &= r^k(s_i^k, a_i^k) \\ &\quad + \zeta Q'_i(s_i^{k+1}, a_i^{k+1}; \theta_{q'_i})|_{a_i^{k+1} = \mu'(s_i^{k+1}; \theta_{\mu'})} \end{aligned} \quad (8.29)$$

19:             Update the actor policy using the sampled policy gradient: $\nabla_{\theta_{\mu_i}} J_i \approx$

$$\frac{1}{K_i} \sum \nabla_{a_i^k} Q_i(s_i^k, a_i^k; \theta_{q_i})|_{a_i^k = \mu_i(s_i^k)} \nabla_{\theta_{\mu_i}} \mu(s_i^k; \theta_{\mu_i}) \quad (8.30)$$

20:             Update the target networks:

$$\theta_{q'_i} \leftarrow \tau \theta_{q_i} + (1 - \tau) \theta_{q'_i} \quad (8.31)$$

$$\theta_{\mu'_i} \leftarrow \tau \theta_{\mu_i} + (1 - \tau) \theta_{\mu'_i} \quad (8.32)$$

21:             Update the state $s_i^t = s_i^{t+1}$
22:         **end for**
23:     **end for**
24: **end for**

encoding has some issues, such as the more data is needed to train the model effectively and the more parameters, the more computation is required to train and use the model; herein, it turns out that making a model more difficult to learn effectively and it is easy to exceed the capabilities of the hardware.

Embedding is rising as a potential technique in which a lower-dimensional space can be achieved by translating from a large sparse vector while preserving semantic relationships [34] to deal with these problems. To apply the embedding layer efficiently in our problem, we divide the input of the $i$th U2U pair into two parts, ID $i$ and QoS constraint $\mathscr{I}_i$. Depending on the number of U2U pairs in the network, the output dimension of embedding layers can be chosen flexibly to reduce the memory storage and processing time while ensuring the performance of the network. The ID $i$ of the U2U pair is put into the embedding layer and a fully connected layer before being concatenated with the level interference of the $i$th U2U pair, $\mathscr{I}_i$. We assume that the concatenated layer is the input of neural networks in the DDPG algorithm. The proposed model's actor and critic network is described in Figure 8.2.

The details of the SDDPG algorithm-based approach for multi-agent power allocation problems in U2U-based communications are described in Algorithm 8.2.

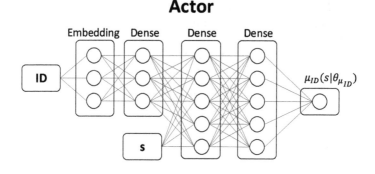

*Figure 8.2* Proposed model with sharing actor and critic network for multi-agent deep reinforcement learning problem

**Algorithm 8.2** Sharing deep deterministic policy gradient for multi-agent power allocation problem in U2U-based communications

1: Initialisation:
2: Initialise the critic network $Q(s, a; \theta_q)$ and actor network $\mu(s; \theta_\mu)$ with random parameter $\theta_q$ and $\theta_\mu$
3: Initialise the target networks $Q'$ and $\mu'$ with parameter $\theta_{q'} \leftarrow \theta_q, \theta_{\mu'} \leftarrow \theta_\mu$
4: Initialise replay buffer $D$
5: **for** U2U $i, i \in N$ **do**
6:    **for** episode = $1, \ldots, M$ **do**
7:       Initialise the embedding layer
8:       Initialise a random process $\mathcal{N}_i$ for action exploration
9:       Receive initial observation state $s_i^1$ by concatenating the output of the embedding layers and $\mathcal{I}_i$
10:       **for** iteration = $1, \ldots, T$ **do**
11:          Obtain the action $a_i^t$ at state $s_i^t$ according to the current policy and exploration noise
12:          Measure the achieved SINR at the receiver according to (8.2)
13:          Update the reward $r_i^t$ according to (8.11)
14:          Observe the new state $s_i^{t+1}$
15:          Store transition $(s_i^t, a_i^t, r_i^t, s_i^{t+1})$ into replay buffer $D$
16:          Sample randomly a mini-batch of $K$ transitions $(s_i^k, a_i^k, r_i^k, s_i^{k+1})$ from buffer $D$
17:          Update critic by minimising the loss:

$$L = \frac{1}{K} \sum \left(y^k - Q(s_i^k, a_i^k; \theta_q)\right)^2 \tag{8.33}$$

where

$$\begin{aligned} y^k = &r(s_i^k, a_i^k) \\ &+ \zeta Q'(s_i^{k+1}, a_i^{k+1}; \theta_{q'})|_{a_i^{k+1} = \mu'(s_i^{k+1}; \theta_{\mu'})} \end{aligned} \tag{8.34}$$

18:          Update the actor policy using the sampled policy gradient: $\nabla_{\theta_\mu} J \approx$

$$\frac{1}{K} \sum \nabla_{a_i^k} Q(s_i^k, a_i^k; \theta_q)|_{a_i^k = \mu_i(s_i^k)} \nabla_{\theta_{\mu_i}} \mu(s_i^k; \theta_{\mu_i}) \tag{8.35}$$

19:          Update the target networks:

$$\theta_{q'} \leftarrow \tau \theta_q + (1 - \tau)\theta_{q'} \tag{8.36}$$

$$\theta_{\mu'} \leftarrow \tau \theta_\mu + (1 - \tau)\theta_{\mu'} \tag{8.37}$$

20:          Update the state $s_i^t = s_i^{t+1}$
21:       **end for**
22:    **end for**
23: **end for**

## 8.5 Simulation results

In this section, we perform the simulation results on PC Intel(R) Core(TM) i7-8700 CPU @ 3.20Ghz to demonstrate the effectiveness of the proposed methods in solving the power control problem in U2U-based communications. Tensorflow version 1.13.1 [35] implements all algorithms. We design the actor and critic networks with one input layer, one output layer, one hidden layer of 100 units, and Adam optimisation algorithm [36] for training. The parameters of neural networks are initialised with small random values with a zero-mean Gaussian distribution. The other simulation parameters are given in Table 8.1.

Figure 8.3 illustrates the EE performance of the network using the DDDPG algorithm while considering different values of mini-batch size $K$ and the learning rate of actor and critic network, $\alpha_A$ and $\alpha_C$, respectively. From Figure 8.3(a), we can see that with a small batch size, our proposed algorithms can be needed to take a long time to reach the optimal policy. On the other hand, there is a possibility that the learning process can be trapped in the local optimum and cannot escape to reach the best performance if we choose too large a batch size, although the calculated gradient is more accurate than the ones with a small batch size; hence, it may lead to a slower convergence. Meanwhile, the parameters of neural networks are updated according to the value of the learning rate. The learning rate decides the speed of convergence and stability of our proposed algorithms. In Figure 8.3(b), with the small values of the learning rate, results are at a slower speed of convergence. On the contrary, if we choose a high learning rate, the algorithms can diverge from the optimal solution. Clearly, our proposed algorithms can achieve the best performance with the learning rate, $\alpha_A = 0.0001$ and $\alpha_C = 0.0001$. Based on the result shown in Figure 8.3, we choose the batch size to be $K = 32$ and the initial learning rate $\alpha = 0.0001$ for actor and critic networks.

Table 8.1 Simulation parameters

| Parameters | Value |
| --- | --- |
| Bandwidth ($W$) | 1 MHz |
| Path-loss exponent | $\alpha_h = 3$ |
| Maximum U2U transmit power | $p_{max} = 23$ dBm |
| Minimum U2U transmit power | $p_{min} = 0$ dBm |
| Channel power gain at the reference | $\beta_0 = -30$ dB |
| Noise power density | $\eta = 0.5$ |
| U2U connection SINR QoS constraint | $\gamma^* = 0$ dB |
| Actor-network learning rate | $\alpha_A = 0.0001$ |
| Critic network learning rate | $\alpha_C = 0.0001$ |
| Soft replacement | $\tau = 0.01$ |
| Discount factor | $\zeta = 0.9$ |
| Memory pool capacity | $D = 10,000$ |
| Mini-batch size | $K = 32$ |

*DDDPG in UAV-to-UAV-based communications* 111

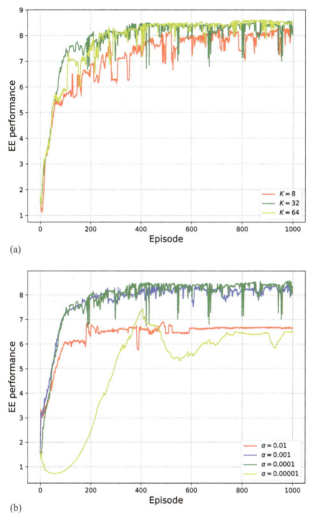

*Figure 8.3  The EE performance of the network by using the DDDPG algorithm in multi-agent power allocation problem in U2U-based communications with different values of batch size K and learning rate $\alpha$, the number of U2U pairs, N = 30*

Figure. 8.4 compares the performance of our two proposed approaches based on the DDDPG and SDDPG algorithm with the output dimension of the embedding layer set to 5, i.e. $|Dims| = 5$. The comparison is against the standard DDPG algorithm for a multi-agent power allocation problem in U2U-based communications. The EE performance of the network when using the DDDPG and SDDPG algorithms is almost identical and better than the standard DDPG in multi-agent learning. In convergence, the speed of convergence with the SDDPG algorithm is faster than that with the DDDPG algorithm and with the standard DDPG algorithm. The reason is

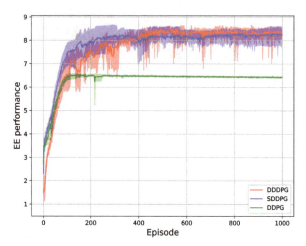

*Figure 8.4  The EE performance of the network by using the DDDPG, SDDPG, and DDPG algorithm in multi-agent power allocation problem in U2U-based communications with the number of U2U pairs, $N = 30$*

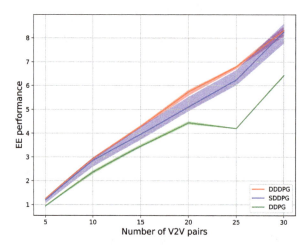

*Figure 8.5  Performance results of the DDDPG, SDDPG, and DDPG algorithm-based approaches with different number of U2U pairs in the network*

that when we use sharing networks for $N$ agents, these networks are trained many times, and the next agents can use the previously pre-trained networks to achieve an optimal policy faster than the DDDPG algorithm-based approach. These results advise that using the combination of multi-agent learning and the DDPG algorithm significantly helps to find an optimal policy for non-cooperative energy-efficient power allocation problems in U2U-based communications.

Figure 8.5 plots the EE results of the network with different numbers of U2U pairs by using the DDDPG, SDDPG, and standard DDPG algorithms. The output dimensions of the embedding layer in the SDDPG algorithm for $N = 5$, $N = 10$, $N = 15$, $N = 20$, $N = 25$, $N = 30$ are $Dims = 2$, $Dims = 3$, $Dims = 3$, $Dims = 4$, $Dims = 4$, and $Dims = 5$, respectively. The performance of the network by using the DDDPG and SDDPG algorithm-based approaches outperforms the ones based on the classical DDPG algorithm in different numbers of U2U pairs. The simulation result difference between models based on the DDDPG and SDDPG algorithm is small even when the number of U2U pairs increases. With $N = 30$, the average performances of the DDDPG and SDDPG algorithm-based approaches are almost identical. In some cases, the scheme's performance based on the SDDPG algorithm is better than that of the DDDPG algorithm. However, the DP algorithm uses $N$ neural networks for the actor function and $N$ neural networks for the critic function. Meanwhile, in the SDDPG algorithm, we share one actor network and one critic network for all the agents. Therefore, the computational processing and memory storage used for the DDDPG algorithm-based approach is many times higher than the SDDPG algorithm when the number of U2U pairs increases.

Next, we compare EE performance results of the network using the SDDPG algorithm-based approach in different output dimensions of the embedding layer in Figure 8.6. With the number of U2U pairs in the network $N = 30$, we can achieve the best performance while setting the output dimension of the embedding layer to $Dims = 5$. The higher output dimensions in the embedding layer do not guarantee better performance. However, the variance of models with the higher output dimensions of the embedding layer is lower.

Moreover, in Figure 8.7, we present the performance of the network with different values of SINR requirements $\gamma^*$ in models using the DDDPG, SDDPG, and classical DDPG algorithm. As we can see from Figure 8.7, when the value of SINR

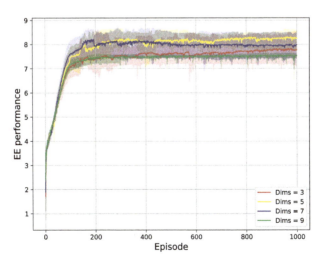

*Figure 8.6*    *The EE performance of the network by using the SDDPG algorithm with different output dimensions of the embedding layer in multi-agent power allocation problem in U2U-based communications with the number of U2U pairs, $N = 30$*

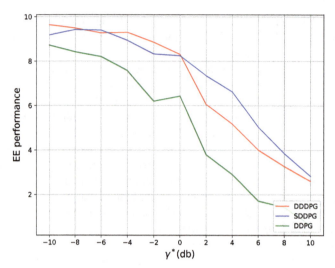

*Figure 8.7* *The EE performance of the network by using the DDDPG, SDDPG, and DDPG algorithm in multi-agent power allocation problem in U2U-based communications while considering different values of SINR requirement, $\gamma^*$*

Table 8.2   Comparison of running time

|  | DDDPG | SDDPG | DDPG | OPA [12] |
| --- | --- | --- | --- | --- |
| $N = 5$ | 4.16 ms | 5.78 ms | 4.12 ms | 121.1 ms |
| $N = 15$ | 8.05 ms | 11.61 ms | 7.36 ms | 130.8 ms |
| $N = 30$ | 21.1 ms | 28.07 ms | 19.89 ms | 145.6 ms |

requirement $\gamma^*$ is too high, the EE result degrades due to the decrease in the number of U2U links that satisfy QoS requirements. The performance of the SDDPG algorithm-based approach is better than the ones using the DDDPG algorithm when we choose the $\gamma^*$ high. In addition, the effectiveness of our proposed algorithms, SDDPG and DDDPG, is superior to the classical DDPG algorithm for multi-agent power control problems in U2U-based communications.

Finally, we evaluate the processing time of the proposed models during test time after training neural networks compared with other approaches. Table 8.2 presents the average processing time in different scenarios. Using our proposed models, each U2U user can choose the power level to maximise the EE performance of the network in milliseconds while satisfying QoS requirements. We only need 21.1 ms and 28.07 ms to solve the power allocation problem in U2U-based communications with the number of U2U pairs, $N = 30$, by using the DDDPG and SDDPG algorithms, respectively. On the other hand, the method based on the logarithmic inequality algorithm in [12] needs 145.6 ms to solve a similar problem with the same environment parameters. Therefore, the results suggest that our proposed models are promising techniques for real-time scenarios.

## 8.6 Summary

This chapter illustrated how to solve a multi-agent energy-efficient power allocation problem in U2U-based communications by adopting two DRL models based on the DDDPG and SDDPG algorithms. The proposed model was able to overcome the limitations of existing approaches. Indeed, by utilising the advantage of neural networks and the embedding layer, simulation results show how it outperformed other baseline algorithms in terms of the EE performance of the network, computational complexity, and memory storage. Last but not least, it is worth mentioning that the solution's computational complexity and memory storage can be significantly reduced by using the SDDPG algorithm when the number of U2U pairs increases. Future research directions include investigating more efficient multi-agent learning approaches and more advanced DL models to improve the learning convergence, reduce the training variance, and reduce the algorithm's computational complexity.

## References

[1] Nguyen KK, Duong TQ, Vien NA, *et al.* Distributed deep deterministic policy gradient for power allocation control in D2D-based V2V communications. *IEEE Access.* 2019;7:164533–164543.

[2] Masaracchia A, Li Y, Nguyen KK, *et al.* UAV-enabled ultra-reliable low-latency communications for 6G: a comprehensive survey. *IEEE Access.* 2021;9:137338–137352.

[3] Kai C, Li H, Xu L, *et al.* Energy-efficient device-to-device communications for green smart cities. *IEEE Trans Ind Inform.* 2018;14(4):1542–1551.

[4] Vo NS, Duong TQ, Tuan HD, *et al.* Optimal video streaming in dense 5G networks with D2D Communications. *IEEE Access.* 2017;6:209–223.

[5] Masaracchia A, Nguyen LD, Duong TQ, *et al.* An energy-efficient clustering and routing framework for disaster relief network. *IEEE Access.* 2019;7:56520–56532.

[6] Asadi A, Wang Q, and Mancuso V. A survey on device-to-device communication in cellular networks. *IEEE Commun Surveys Tut.* 2014 Fourthquarter;16(4):1801–1819.

[7] Sun W, Yuan D, Ström EG, *et al.* Cluster-based radio resource management for D2D-supported safety-critical V2X communications. *IEEE Trans Wireless Commun.* 2016;15(4):2756–2769.

[8] Ren Y, Liu F, Liu Z, *et al.* Power control in D2D-based vehicular communication networks. *IEEE Trans Veh Technol.* 2015;64(12):5547–5562.

[9] Sun W, Ström EG, Brännström F, *et al.* Radio resource management for D2D-based V2V communication. *IEEE Trans Veh Technol.* 2016;65(8):6636–6650.

[10] Sheng Z, Tuan HD, Nasir AA, et al. Power allocation for energy efficiency and secrecy of wireless interference networks. *IEEE Trans Wireless Commun.* 2018;17(6):3737–3751.

[11] Gao H, Zhang S, Su Y, et al. Joint resource allocation and power control algorithm for cooperative D2D heterogeneous networks. *IEEE Access.* 2019;7:20632–20643.

[12] Nguyen MN, Nguyen LD, Duong TQ, et al. Real-time optimal resource allocation for embedded UAV communication systems. *IEEE Wireless Commun Lett.* 2019;8(1):225–228.

[13] Lee J and Lee JH. Performance analysis and resource allocation for cooperative D2D communication in cellular networks with multiple D2D pairs. *IEEE Commun Lett.* 2019;23(5):909–912.

[14] Liu S, Wu Y, Li L, et al. A two-stage energy-efficient approach for joint power control and channel allocation in D2D communication. *IEEE Access.* 2019;7:16940–16951.

[15] Zhang P, Gao AY, and Theel O. Bandit learning with concurrent transmissions for energy-efficient flooding in sensor networks. *EAI Endorsed Transactions on Industrial Networks and Intelligent Systems.* 2018;4(13):1–14.

[16] Nguyen LD, Kortun A, and Duong TQ. An introduction of real-time embedded optimisation programming for UAV systems under disaster communication. *EAI Endorsed Transactions on Industrial Networks and Intelligent Systems.* 2018;5(17):1–8.

[17] Nguyen MT, Nguyen HM, Masaracchia A, et al. Stochastic-based power consumption analysis for data transmission in wireless sensor networks. *EAI Endorsed Transactions on Industrial Networks and Intelligent Systems.* 2019;6(19):1–11.

[18] Ortiz A, Al-Shatri H, Li X, et al. Reinforcement learning for energy harvesting decode-and-forward two-hop communications. *IEEE Trans Green Commun Netw.* 2017;1(3):309–319.

[19] Liu M, Song T, and Gui G. Deep cognitive perspective: resource allocation for NOMA based heterogeneous IoT with imperfect SIC. *IEEE Internet Things J.* 2019;6(2):2885–2894.

[20] Liu M, Song T, Hu J, et al. Deep learning-inspired message passing algorithm for efficient resource allocation in cognitive radio networks. *IEEE Trans Veh Technol.* 2019;68(1):641–653.

[21] Wang Y, Liu M, Yang J, et al. Data-driven deep learning for automatic modulation recognition in cognitive radios. *IEEE Trans Veh Technol.* 2019;68(4):4074–4077.

[22] Huang H, Xia W, Xiong J, et al. Unsupervised learning-based fast beamforming design for downlink MIMO. *IEEE Access.* 2019;7:7599–7605.

[23] Huang H, Song Y, Yang J, et al. Deep-learning-based millimeter-wave massive MIMO for hybrid precoding. *IEEE Trans Veh Technol.* 2019;68(3):3027–3032.

[24] Gui G, Huang H, Song Y, et al. Deep learning for an effective nonorthogonal multiple access scheme. *IEEE Trans Veh Technol.* 2018;67(9):8440–8450.

[25] Huang H, Guo S, Gui G, *et al. Deep learning for physical-layer 5G wireless techniques: opportunities, challenges and solutions*; 2019. Available from: arXivpreprintarXiv:1904.09673.

[26] Nguyen KK, Duong TQ, Vien NA, *et al.* Non-cooperative energy efficient power allocation game in D2D communication: a multi-agent deep reinforcement learning approach. *IEEE Access*. 2019;7:100480–100490.

[27] Ye H, Li GY, and Juang BHF. Deep reinforcement learning based resource allocation for V2V communications. *IEEE Trans Veh Technol*. 2019;68(4): 3163–3173.

[28] T. P. Lillicrap, J. J. Hunt, A. Pritzel *et al.* "Continuous control with deep reinforcement learning," in *Proc. 4th International Conf. on Learning Representations (ICLR)*, 2016.

[29] Puterman ML. *Markov Decision Processes: Discrete Stochastic Dynamic Programming*. John Wiley & Sons, Inc.; 1994.

[30] Arulkumaran K, Deisenroth MP, Brundage M, *et al.* Deep reinforcement learning: a brief survey. *IEEE Signal Process Mag*. 2017;34(6):26–38.

[31] Bertsekas DP. *Dynamic Programming and Optimal Control*. vol. 1. Athena Scientific Belmont, MA; 1995.

[32] Sutton RS, McAllester D, Singh S, *et al.* Policy gradient methods for reinforcement learning with function approximation. In: *Adv. Neural Inf. Process. Syst.*; 2000. p. 1057–1063.

[33] Gupta JK, Egorov M, and Kochenderfer M. Cooperative multi-agent control using deep reinforcement learning. In: *Int. Conf. Autonomous Agents Multiagent Syst.*; 2017. p. 66–83.

[34] Gal Y and Ghahramani Z. A theoretically grounded application of dropout in recurrent neural networks. In: *Adv. Neural Inf. Process. Syst.*; 2016. p. 1019–1027.

[35] M. Abadi, P. Barham, J. Chen, Z. Chen, A. Davis, J. Dean, *et al.* "Tensorflow: A system for largescale machine learning," in *Proc. 12th USENIX Sym. Opr. Syst. Design and Imp. (OSDI 16)*, Nov. 2016, pp. 265–283.

[36] Kingma DP and Ba JL. *Adam: a method for stochastic optimization*; 2014. Available from: https://arxiv.org/abs/1412.6980.

*Chapter 9*
# Non-cooperative energy-efficient power allocation game in UAV-to-UAV communication: a multi-agent deep reinforcement learning approach

As already mentioned in previous chapters, the recent widespread of mobile devices and sensors, as well as the rapid emergence of new wireless and networking technologies such as wireless sensor networks, UAV-to-UAV (U2U) communication, and vehicular ad hoc networks, are expected to achieve a considerable increase in data rates, coverage and the number of connected devices. At the same time, there will be the need for reduced communication latency and energy consumption for such energy-constrained devices, calling them for more advanced techniques to achieve an optimal trade-off between energy consumption and achievable network performance. This chapter illustrates how using RL represents an efficient optimisation framework to tackle this problem in real time. More specifically, the illustrated approach is based on DRL to deal with the energy-efficient power allocation problem while still satisfying the quality of service constraints in U2U communication.*

## 9.1 Introduction

With a fast-growing number of mobile devices and sensors, wireless networks (e.g. heterogeneous networks (HetNets), ultra-dense networks, and unmanned aerial vehicle (UAV) networks) become more autonomous, complex and dynamic in nature. At the same time, it incurs a fast escalation of energy demand and requirements for efficient resource allocation. In other words, the critical problem for wireless network energy efficiency is the trade-off between energy consumption and guaranteed performance (e.g. throughput or quality of service (QoS)). Technologies towards energy-efficient wireless networks have established a long research history in which most studies are focused on energy-efficient data communications [2]. To obtain better performance, a substantial effort has been spent on optimisation theory to develop more efficient algorithms to gain optimal or near-optimal solutions. However, most previous work assumes a static network environment. The environment is

---

*This chapter has been published partly in [1].

often dynamically unstable in advanced wireless networks with an enormous number of devices. Therefore, enabling network nodes to make autonomous decisions is desirable. The decisions must be based directly on local observations, e.g. power allocation, spectrum access, and interference management, to maximise the network performance.

Reinforcement learning (RL) [3] is a sub-field of ML which offers a mathematically principled framework studying how an autonomous agent makes optimal sequential decisions. An RL-based agent learns to make optimal decisions through trial-and-error interactions directly with a dynamic environment in which the objective is to maximise the task's performance measure, e.g. user experience. In an interactive learning fashion, the agent takes actions and observes the results of the interactions to make decisions. Over the last 20 years, there have been many successful applications of RL ranging from robotics [4], natural language processing [5] to game solving [6]. One of the most famous applications of RL is AlphaGo [7], the first ever computer program which can beat a world-class professional player in the *Go* board game.

In the area of wireless communications, RL has recently emerged as a powerful tool for dealing with problems in modern networks. However, under the uncertain, stochastic, and large-scale environment, the computational complexity of existing techniques grows exponentially and becomes unmanageable. This makes existing learning approaches slower to converge as the agent has to explore the whole state space repeatedly to search for the best policy for an entire system. Consequently, standard RL approaches become inefficient and impractical in complex domains. Recently, DL [8] techniques have brought revolutionary advances in computer vision. DL is a sub-field of ML concerned with flexible and scalable algorithms to optimise complex functions on a large dataset. Over the past few years, DL has been applied successfully in many areas, for example, face recognition [9–12], cyber security [13], and game playing [7]. Combining DL and RL leads to a new paradigm called deep reinforcement learning (DRL) [14].

This chapter proposes to apply DRL to optimise power allocation in U2U communication. In particular, we develop single-agent RL and multi-agent RL that are able to obtain an optimal policy. We then propose three different low-complexity algorithms, which are based on (i) deep Q-learning, (ii) double deep Q-learning, and (iii) dueling deep Q-learning. These proposed approaches can solve the noncooperative game in U2U communication in milliseconds. Our major contributions to this chapter can be summarised as follows:

- We propose a novel method based on DRL for power allocation problems in U2U communication. Based on this method, each U2U transmitter can optimise its power used for information transmission to adapt to the dynamics of the environment. Our method will be based on deep Q-networks, double deep Q-networks, and dueling deep Q-networks algorithms.
- We directly tackle the non-cooperative problem in U2U communication. We assume that each U2U pair in the network is parsimonious and unaware of other U2U pairs' power allocation schemes and conditions. After a certain period, all

U2U pairs broadcast their strategies to peers to calculate the overall performance of the network.
- We perform extensive experiments with the aim not only to demonstrate the efficiency of the proposed solution in comparison with other conventional methods but also to provide running time results in detail.

The reminder of this chapter is organised as follows. Section 9.2 discusses related work using RL and DL for physical layer and resource allocation. The system model and the resource allocation problem in U2U communications are described in Section 9.3. Section 9.4 describes single-agent RL and multi-agent RL. In Section 9.5, we show how we adapt existing RL algorithms to solve the non-cooperative resource allocation problem. In Section 9.6, we present numerical simulations to verify the proposed schemes. Finally, Section 9.7 concludes this chapter.

## 9.2 Related work

Energy-efficient power allocation is a challenging problem in modern wireless networks with many QoS constraints. In particular, computational processing and limited environmental knowledge always restrict U2U-like device-to-device (D2D) communication. Several solutions have been proposed to tackle these problems [15–19]. In [15], the author proposed a joint link scheduling and power allocation optimisation algorithm to maximise the system throughput of D2D-assisted wireless caching networks. However, this work assumes that the network has a single channel to simplify the analysis and implementation. In addition, it assumes a centralised control mechanism; therefore, it can only result in sub-optimal solutions. Moreover, this method requires all data to be collected by the base station, thus introducing a time delay in the network. An alternative algorithm is based on the logarithmic inequality to find an optimal resource allocation strategy for UAV communication systems [18]. This algorithm shows promising performance regarding both the sum rate and running time. However, it requires all data to be processed in a centralised mode, therefore the running time increases rapidly as the number of D2D pairs increases. Very recently, real-time optimisation has been considered to increase the running time of optimisation for radio resources (see, e.g. [18,20] and references therein).

Over the past few years, RL has been applied widely in wireless networks to allow each node to possess its own self-organising function, which leads to a distributed control mode. For example, the work in [21] proposes a cooperative Q-learning method to deal with a resource allocation problem in heterogeneous wireless networks to maximise the sum capacity of the network while being able to ensure QoS and fairness to users. Regarding problems in D2D communication, the work in [22] proposes two approaches using RL: convergence-based and epsilon-greedy Q-learning. They aim to find an optimal energy-efficient power allocation policy for the energy harvesting problem.

With the emergence of neural networks, DL algorithms have received considerable attention in the field of wireless communications [23–29] because they can offer a powerful optimisation tool for non-convex, non-differentiable and complex objective functions. It works by constructing a very big neural network and adjusting its weights using stochastic gradient descent, i.e. using *mini-batch* (a small subset of all available data). In the resource allocation problem, DL combined with damped three-dimensional message-passing can be used to minimise the weighted sum of the secondary interference power in cognitive radio networks [28]. Another work using deep recurrent neural networks [29] proposed to look for an optimal resource allocation scheme for the mobile users and the IoT users in the non-orthogonal multiple access (NOMA)-based heterogeneous IoT. These DL-based algorithms are shown to achieve optimal solutions with a reasonably fast convergence rate and low computational complexity. DL has recently become a popular research topic in the physical layer as well [27]. The work in [23] used convolutional neural networks to improve the accuracy of automatic modulation recognition in the cognitive radio. Deep neural networks have also been applied to solve many existing problems in massive multiple-input multiple-output (MIMO) very efficiently [24,25]. Long short-term memory networks are also used in [26] to detect channel characteristics automatically. After being optimised offline, the trained neural network-aided NOMA system can achieve much-improved performance in terms of the error rate and sum data rate. In comparison to conventional approaches, DL-based methods have shown better performance and improved computation in many domains, such as beamforming design methods and hybrid precoding. In addition, it shows great potential for applications in real-time scenarios.

In recent years, DRL has received much interest as an alternative and powerful tool to deal with complex optimisation problems in wireless networks [30,31]. The work in [30] proposed a DRL-based joint mode selection and resource management approach to optimise the system power consumption for green fog radio access networks. Each user equipment is assumed to operate in either cloud RAN or D2D mode. The resources to be managed include both the radio resource and the computing resource. Through DRL, this work shows that the network controller can be optimised to make optimal decisions under the dynamics of edge caching states.

DRL is also applied in vehicle-to-vehicle communication (V2V) [31] to find an optimal sub-band and power level for transmission in every link. Each V2V link is considered an RL agent in which the spectrum and transmission power are considered as actions and optimised depending on all information as environment observations such as local channel state information, interference levels, the resource allocation decision of instantaneous channel conditions, and exchanged information shared from the neighbours at each time slot. In principle, the RL agent learns autonomously how to balance between minimising the interference of V2V links to the vehicle-to-infrastructure networks and satisfying the stringent latency constraints imposed on V2V links.

However, most existing approaches assume data samples are independently and identically distributed. Furthermore, each node is assumed to know about other nodes' condition and their resource allocation strategies. Moreover, these approaches are still non-trivial in scaling to real-time scenarios due to the high computational complexity. In this chapter, we present a non-cooperative energy-efficient approach based on DRL. By utilising the advantages of DL, the proposed algorithms can solve the non-cooperative game and enable each U2U transmitter to have a self-organising function to adapt to the dynamic environment in milliseconds.

## 9.3 System model and problem formulation

We consider a communication system that includes $N$ U2U pairs, where the U2U transmitters (U2U-Tx) and U2U receivers (U2U-Rx) are randomly distributed within the coverage of one base station (BS). Each U2U-Tx/U2U-Rx is equipped with a single antenna. We denote U2U($i$) as the $i$th U2U pair, where $i \in \{1, 2, \ldots, N\}$.

We assume that the location of U2U-Tx and U2U-Rx are $(x_{\text{Tx}}^n, y_{\text{Tx}}^n)$ and $(x_{\text{Rx}}^n, y_{\text{Rx}}^n)$, respectively. The distance between the $i$th U2U-Tx and U2U-Rx pair, which are assumed to be at the same altitude $H_z$, is calculated using Euclidean distance as:

$$R_i = \sqrt{(x_{\text{Tx}}^i - x_{\text{Rx}}^i)^2 + (y_{\text{Tx}}^i - y_{\text{Rx}}^i)^2}. \tag{9.1}$$

The channel power gain between U2U-Tx and U2U-Rx is defined as:

$$h_{ii} = \beta_0 p_i^2 R_i^{-\alpha_h}, \tag{9.2}$$

where $\beta_0$ is the channel power gain at the reference distance $d_0$, $p_i$ is an exponentially distributed random variable with unit mean, $R_i$ is the distance between the $i$th U2U-Tx and U2U-Rx pair, and $\alpha_h$ is the path-loss exponent for U2U links.

The total interference plus noise at each U2U user includes the interferences from all U2U-Txs and the additive white Gaussian noise (AWGN). We use $p_i$ ($p_i^{\min} \leq p_i \leq p_i^{\max}$) and $\gamma_i$ to designate the transmission power and received signal-to-interference-plus-noise ratio (SINR) at U2U $i$, respectively. The SINR at the $i$th U2U user is written as:

$$\gamma_i = \frac{p_i h_{ii}}{\sum_{j \in N}^{j \neq i} p_j h_{ji} + \sigma^2}, \tag{9.3}$$

where $\sigma$ is the AWGN's power.

The goal of power allocation is to ensure that no SINR falls below its threshold $\gamma_i^*$ which is chosen to guarantee the QoS constraint as:

$$\gamma_i \geq \gamma_i^*, \quad \forall i \in N. \tag{9.4}$$

We focus on a system performance that encompasses the U2U pair capacity. This suggests us to define a reward function that can be used in an RL algorithm later as follows:

$$\mathscr{R}_i = \frac{W \ln(1 + \gamma_i)}{p_i}, \tag{9.5}$$

where $W$ is a bandwidth. The reward of each U2U user is a function of the joint actions of all U2U pairs.

Formally, the power allocation game can be defined as a constrained optimisation problem as follows:

$$\max \sum_{i}^{N} \frac{W}{p_i} \ln\left(1 + \frac{p_i h_{ii}}{\sum_{j \in N}^{j \neq i} p_j h_{ji} + \sigma^2}\right), \quad (9.6)$$

$$\text{s.t} \quad \gamma_i \geq \gamma^*, \forall i \in N, \quad (9.7)$$

$$p_i^{\min} \leq p_i \leq p_i^{\max}. \quad (9.8)$$

There are some challenges in solving the above optimisation problem because of the fact that every U2U user does not know the power allocation strategies in other U2U pairs. The U2U user can only obtain local information such as environment state and power allocation scheme. Therefore, we propose a multi-agent RL approach to find an optimal power allocation scheme for each U2U user.

## 9.4 Reinforcement learning for energy-efficient power allocation game in U2U communication

In this section, we discuss the background of single-agent RL and multi-agent RL.

### 9.4.1 Single-agent Q-learning

We assume there is an agent whose task is to find an optimal policy through interactions with an environment. This problem can be formulated as a Markov decision process (MDP) defined as a 4-tuple $\langle \mathcal{S}, \mathcal{A}, \mathcal{R}, \mathcal{P} \rangle$, where $\mathcal{S} = \{s_1, s_2, \cdots, s_m\}$ is the finite set of state, $\mathcal{A} = \{a_1, a_2, \cdots, a_l\}$ is a set of agent discrete actions. The reward function $r = \mathcal{R}(s, a, s')$ is defined at state $s \in \mathcal{S}$, action $a \in \mathcal{A}$, and next state $s'$. The transition function $\mathcal{P}_{ss'(a)} = p(s'|s, a)$ defines the probability of transitioning to state $s'$ if the agent is at state $s$ and takes action $a$.

In RL, the goal is to find the optimal policy $\pi^*(s)$ for each $s$, which maximises the total expected discounted reward. Under the policy, $\pi$, the value of a state is defined as:

$$V^\pi(s) = \mathbb{E}\left\{\sum_{k=0}^{\infty} \gamma^k r_i^{k+1} \Big| s^0 = s\right\}, \quad (9.9)$$

where $0 \leq \gamma \leq 1$ is the discount factor, and $\mathbb{E}\{\cdot\}$ denotes the expectation operation, which is related to the stochastic property of the policy $\pi$ and dynamics $\mathcal{P}_{ss'}(a)$. We can rewrite (9.9) to the Bellman equation as:

$$V^\pi(s) = \mathbb{E}\{r_t | s_t = s\} + \gamma \sum_{s' \in \mathcal{S}} \mathcal{P}_{ss'}(a) V^\pi(s'). \quad (9.10)$$

The optimal policy $\pi^*$, which is a mapping from the states to the optimal actions, would maximise the expected cumulative reward. There is at least one optimal strategy $\pi^*$ that satisfies the following Bellman equation [32]:

$$V^*(s) = V^{\pi^*} = \max_{a \in \mathscr{A}}\{\mathbb{E}(r(s,a)) + \gamma \sum_{s' \in \mathscr{S}} P_{ss'} V^*(s')\}. \tag{9.11}$$

The action-value function is the expected reward starting from state $s$, taking action $a$ under the policy $\pi$:

$$Q^\pi(s,a) = \mathbb{E}(r(s,a)) + \gamma \sum_{s' \in \mathscr{S}} P_{ss'} V(s'). \tag{9.12}$$

The optimal policy $Q^*(s,a)$ is defined as:

$$Q^*(s,a) = Q^{\pi^*}. \tag{9.13}$$

Then we get:

$$V^*(s) = \max_{a \in \mathscr{A}} Q^*(s,a). \tag{9.14}$$

Q-learning [3] is one of RL algorithms that tries to learn $\pi^*$ without knowing the environment dynamics $\mathscr{P}_{ss'}(a)$. The agent learns by adjusting $Q$ value according to the update rule:

$$Q^t(s,a) = (1-\alpha)Q^t(s,a) + \alpha[r^t + \gamma \max_{a' \in \mathscr{A}} Q^t(s',a')], \tag{9.15}$$

where $\alpha \in [0,1)$ is the learning rate.

### 9.4.2 Multi-agent Q-learning approach

Our goal is to find a strategy to maximise the energy efficiency of a U2U communication system when the whole network consists of a large number of U2U users. Hence, we define U2U-Tx in each U2U pair as one RL agent as introduced in 9.4.1, which just only knows the information of its own environment. The network can be considered as a multi-agent system in which the optimal scheme consists of the optimal policies of individual agents. We now describe *multi-agent Q-learning algorithm* that is expected to solve the complex power control issue in this setting.

Multi-agent Q-learning has $N$ agents corresponding to $N$ U2U pairs; each agent is equipped with a classical Q-learning algorithm and learns without cooperating with the other agents. The $i$th agent is represented as a tuple $\langle \mathscr{S}_i, \mathscr{A}_i, \mathscr{R}_i, \mathscr{P}_i \rangle$, where $\mathscr{S}_i$ is a finite set of environment states, $\mathscr{A}_i$ is a finite set of agent actions, $\mathscr{R}_i$ is the reward function, and $\mathscr{P}_i$ is the state transition probability function of the $i$th agent. We define the agents, states, actions and reward functions as follows:

**Agent**: Each agent is one U2U transmitter. The system would consist of $N$ such agents.

**State**: The state of $i$th U2U transmitter at time $t$ is defined as:

$$\mathscr{S}_i^t = (i, \mathscr{I}_i), \tag{9.16}$$

where $\mathscr{I}_t \in (0, 1)$ indicates the level of interference as:

$$\mathscr{I}_i = \begin{cases} 1 & \text{for} \quad \gamma_i \geq \gamma^* \\ 0 & \text{for} \quad \text{otherwise} \end{cases} \quad (9.17)$$

**Action**: The action of each agent consists of a set of transmitting power levels. It is denoted as:

$$\mathscr{A} = (a_1, a_2, ..., a_l), \quad (9.18)$$

where $l$ represents that every agent has $l$ power levels. There are many ways to choose actions based on the current action-value estimation, for example, $\varepsilon$-greedy strategy, which is described as follows:

$$\pi_a^s = \begin{cases} \operatorname{argmax}_{a \in \mathscr{A}} Q(s, a) & \text{with probability} \quad 1 - \varepsilon \\ \text{random action} & \text{with probability} \quad \varepsilon \end{cases} \quad (9.19)$$

**Reward**: The reward $\mathscr{R}_i$ of $i$th U2U user in state $s_i$ is the immediate return due to the execution of action $a_i$:

$$\mathscr{R}_i = \begin{cases} \frac{W \ln(1+\gamma_i)}{p_i} & \text{if} \quad \mathscr{I}_i = 1 \\ 0 & \text{if} \quad \mathscr{I}_i = 0 \end{cases} \quad (9.20)$$

The multi-agent Q-learning algorithm finds optimal Q-value $Q^*(s_i, a_i)$ in a recursive way after receiving a transition information $\langle s_i, a_i, s'_i, \pi_i \rangle$, where $s_i^t = s_i \in \mathscr{S}_i$ and $s_i^{t+1} = s'_i \in \mathscr{S}_i$ are the environment states observed by agent $i$ at time slot $t$ and $t+1$, respectively; $a_i^t = a_i \in \mathscr{A}_i$ and $a_i^{t+1} = a'_i \in \mathscr{A}$ and $\pi_i$ are the $i$th agent' action at time slot $t$ and the transmission strategy during time slot $t$. The update rule for Q-learning is given by $Q_i^{t+1}(s_i, a_i)$:

$$Q_i^t(s_i, a_i) + \alpha^t \{r_i^t + \gamma \max_{a'_i \in \mathscr{A}} Q_i^t(s'_i, a'_i) - Q_i^t(s_i, a_i)\}, \quad (9.21)$$

---

**Algorithm 9.1** Multi-agent Q-learning for power allocation optimisation

1: Initialisation:
2: **for** $i \in N$ **do**
3:     Initialise the value function $Q_i(s, a, s') = 0$
4:     Initialise the initial state $s_i$ according to (9.17)
5: **end for**
6: **while** not convergence **do**
7:     **for** U2U $i, i \in N$ **do**
8:         Select action $a_i^t$ according to the strategy $\pi_i$ (9.19)
9:         Measure SINR at the receiver according to (9.3)
10:        Update the reward $r_i^t$ according to (9.20)
11:        Observe the new state $s_i^{t+1}$
12:        Update the action-value $Q_i(s_i, a_i)$ according to (9.21)
13:        Update the state $s_i^t = s_i^{t+1}$
14:     **end for**
15: **end while**

where $\alpha$ is a learning rate and $\gamma$ is a discount factor. The pseudo-code of the multi-agent power allocation algorithm is depicted in Algorithm 9.1. At each iteration, each agent acts and updates its own policy independently.

## 9.5 Deep reinforcement learning for power allocation optimisation in U2U communication

The approaches based on standard Q-learning as described in 9.4.1 and 9.4.2 can work well when the problem has a small set of states and actions [3]. However, when the problem size increases, it yields many limitations. First, the convergence rate might become slow when the problem is large, and the number of agents $N$ is sufficiently large. Hence, it cannot be adapted to real-time scenarios. Second, the storage of the lookup table $Q_i(s, a, s')$ (for every agent $i$, states $s, s'$, and action $a$) becomes impractical and has limited generalisation ability. These issues can be solved using DRL [33] where both the use of function approximation (i.e. using deep neural networks to approximate $Q_i$) and DL (i.e. training via stochastic gradient descent) play an important role to scale Q-learning up to much larger domains.

In addition, the performance of RL algorithms might become unstable or even diverge when a non-linear function approximation is used. This is due to the fact that a small change of Q-value may result in a big change to the policy. To address this issue, the authors in [33] also propose a deep Q-network (DQN) which is basically a DRL method with two major techniques: *experience replay* mechanism and *target Q-network*:

- *Experience replay mechanism*: The algorithm store transitions $\langle s_t, a_t, r_t, s'_t \rangle$ in a replay memory (*a buffer memory*). At each learning step, a mini-batch is sampled randomly from the memory pool and then a stochastic gradient descent is used to update the weights of the network representing $Q$. By doing so, the previous experiences are exploited more efficiently as the algorithm can re-use them. It ensures the fairness and robustness of DRL. Additionally, by using the experience replay, the data is more likely independent and identically distributed.
- *Target Q-network*: In the training process, the Q-value functions are updated continuously. Thus, the value estimations can become unstable. To address this issue, DQN uses two networks: a target network $\hat{Q}$ and the main network $Q$. The main network is still updated and used, as previously mentioned. $\hat{Q}$ is used to store an old and stable main $Q$ network, which helps slow down the changes. This technique can guarantee performance improvement gradually.

DQN inherits and promotes the advantages of both RL and DL techniques, and thus it has a wide range of applications in practice from resource allocation to interference management and security. We now describe the main DQN algorithm and its two variants: double deep Q-learning (DDQN) and dueling deep Q-learning.

### 9.5.1 Deep Q-learning

We assume the experience replay stores the last $D$ tuples of $\langle s, a, r, s' \rangle$ in a replay buffer. We take mini-batches of random samples from this buffer to train Q-networks to ensure fairness and robustness. In DQN, we use two networks to store the Q-value.

The first one $Q_w$ with weights $w$ is constantly updated while the second one, the target Q-network $\hat{Q}$ with weights $\hat{w}$, is synchronised from the first network once in a while. To estimate the network, we optimise the following sequence of loss functions:

$$L_i = \mathbb{E}\left[(y_i^{DQN} - Q(s_i, a_i; w)^2\right], \qquad (9.22)$$

with

$$y_i^{DQN} = r_i + \gamma \max_{a' \in \mathcal{A}} \hat{Q}(s_i', a'; \hat{w}). \qquad (9.23)$$

During the training stage of the Q-network, instead of using only current experience, the network parameters are updated through training by the random samples mini-batch $\tilde{D}$ from the memory pool $D$ using stochastic gradient descent. In particular, the gradient updating $w$ is computed as:

$$\nabla_w L_i \approx \frac{1}{|\tilde{D}|} \sum_{\{s_i, a_i, r_i, s_i'\} \in \tilde{D}} 2(y_i^{DQN} - Q(s_i, a_i; w)) \nabla_w Q(s_i, a_i; w). \qquad (9.24)$$

Experience replay increases data efficiency through experience reuse in multiple updates. Moreover, uniform sampling from the replay buffer can reduce variance because it can reduce the correlation among the samples. We propose a multi-agent deep Q-learning algorithm for power allocation in U2U communication as described in Algorithm 9.2 where $\tau$ is the network update parameter.

### 9.5.2 Double deep Q-learning

Deep Q-learning is known to learn unrealistically high action values because it includes a maximisation step over estimated action values, which tends to prefer overestimated to underestimated values. The core idea of the double Q-learning algorithm is to reduce overestimation by decomposing the max operation in the target into action selection and action evaluation [33].

The max operation uses the same values to select and evaluate an action in Q-learning and deep Q-learning. This can, therefore, lead to over-optimistic value estimates. To mitigate this problem, deep double Q-learning (DDQL) uses the following target:

$$y_i^{DDQN} = r_i + \gamma Q(s_i', \operatorname*{argmax}_{a' \in \mathcal{A}} \hat{Q}(s_i', a_i')). \qquad (9.25)$$

The details of our proposed multi-agent DDQL for power allocation in U2U communication are provided in Algorithm 9.3.

### 9.5.3 Dueling deep Q-learning

The main idea of dueling deep Q-learning is to reduce the variance in Q-learning, DQN and DDQL. Its main idea originates from a simple technique in statistics in which an estimate bias term can be used to reduce the variance of the Q-value function estimation.

As a result, the deep dueling Q-learning network model [34] consists of two sequences (or streams) of fully connected layers. Both streams are constructed in

**Algorithm 9.2** Multi-agent deep Q-learning for power allocation optimisation with experience replay

1: **Initialisation:**
2: **for** $i \in N$ **do**
3:    Randomly initialise the Q-network $Q_i(s, a; w)$
4:    Randomly initialise the target Q-network $\hat{Q}_i(s, a; \hat{w})$
5:    Initialise the replay memory $D$ with capacity $C$
6: **end for**
7: **while** not convergence **do**
8:    **for** U2U $i, i \in N$ **do**
9:      **for** Iteration **do**
10:        Select action $a_i^t$ using $\varepsilon$-greedy via $Q_i(s_i^t, a_i^t; w)$
11:        Measure SINR at the receiver according to (9.3)
12:        Update the reward $r_i^t$ according to (9.20)
13:        Observe the new state $s_i^{t+1}$
14:        Store transition $(s_i^t, a_i^t, r_i^t, s_i')$ in $D$
15:        **if** the replay memory $D$ is full **then**
16:           Sample a mini-batch of $K$ transitions from $D$
17:           Using stochastic gradient to minimise the loss:

$$\left[ r_i + \gamma \max_{a' \in \mathcal{A}} \hat{Q}_i(s_i', a_i'; \hat{w}) - Q(s_i, a_i; w) \right]^2 \quad (9.26)$$

18:        **end if**
19:        Update the state $s_i^t = s_i^{t+1}$
20:        Update target network: $\hat{w} = \tau \hat{w} + (1 - \tau) w$
21:      **end for**
22:    **end for**
23: **end while**

such a way that they have the capability of providing separate estimates of the value $V(s)$ and advantage $A(s, a)$ functions. Finally, the two streams are combined to produce a single output Q-function $Q(s, a) = A(s, a) + V(s)$. Since the output of the dueling network is a Q-function, it can be trained with many existing algorithms such as DDQL, DQN, and SARSA.

The rationale for the use of the advantage function $A^\pi(s, a) = Q^\pi(s, a) - V^\pi(s)$ is that this allows the agent to determine not just how good its actions are but also how much better their turn out to be than expected. Intuitively, this allows the agent to focus on where the network prediction is lacking. For the state-value function $V^\pi(s) = \mathbb{E}_{a \sim \pi(s)}[Q^\pi(s, a)]$, it follows that $\mathbb{E}_{a \sim \pi(s)[A^\pi(s,a)]} = 0$. In addition, given a deterministic policy, $a^* = \mathrm{argmax}_{a' \in \mathcal{A}} Q(s, a')$, we have $Q(s, a^*) = V(s)$ and hence $A(s, a^*) = 0$.

The function $Q(s, a)$ is only a parameterised estimate of the true Q-function as output. Therefore, there is a lack of identifiability because if given $Q$, we cannot recover $V$ and $A$ uniquely, which leads to poor performance. To address this issue, the authors in [34] propose an objective that forces the advantage function estimator

**Algorithm 9.3** Multi-agent double deep Q-learning for power allocation optimisation with experience replay

1: Initialisation:
2: **for** $i \in N$ **do**
3:     Randomly initialise the Q-network $Q(s, a; w)$
4:     Randomly initialise the target Q-network $\hat{Q}(s, a; \hat{w})$
5:     Initialise the replay memory $D$ with capacity $C$
6: **end for**
7: **while** Not convergence **do**
8:     **for** U2U $i, i \in N$ **do**
9:         **for** Iteration **do**
10:            Select action $a_i^t$ using $\varepsilon$-greedy via $Q_i(s_i^t, a_i^t; w)$
11:            Measure SINR at the receiver according to (9.3)
12:            Update the reward $r_i^t$ according to (9.20)
13:            Observe the new state $s_i^{t+1}$
14:            Store transition $(s_i^t, a_i^t, r_i^t, s_i')$ in $D$
15:            **if** the replay memory D is full **then**
16:                Sample a mini-batch of $K$ transitions from $D$
17:                Using stochastic gradient to minimise the loss:

$$\left[ r_i + \gamma Q_i(s', \underset{a' \in \mathscr{A}}{\operatorname{argmax}} \hat{Q}_i(s_i', a_i')) - Q(s_i, a_i) \right]^2$$

18:            **end if**
19:            Update the state $s_i^t = s_i^{t+1}$
20:            Update target network: $\hat{w} = \tau \hat{w} + (1 - \tau)w$
21:        **end for**
22:    **end for**
23: **end while**

to have zero advantage at the chosen action by letting the last module of the network implement the forward mapping:

$$Q(s, a) = V(s) + \left( A(s, a) - \max_{a' \in |\mathscr{A}|} A(s, a') \right). \quad (9.27)$$

For $a^* = \operatorname{argmax}_{a' \in \mathscr{A}} Q(s, a') = \operatorname{argmax}_{a' \in \mathscr{A}} A(s, a')$, we obtain $Q(s, a^*) = V(s)$. Hence, the stream $V(s)$ provides an estimate of the value function, while the other stream produces an estimate of the advantage function.

To simplify the computation of gradients, (9.27) can be transformed by replacing the max operator with an average as:

$$Q(s, a) = V(s) + \left( A(s, a) - \frac{1}{|\mathscr{A}|} \sum A(s, a) \right). \quad (9.28)$$

The rank of $A$ is not be changed by subtracting the mean in (9.28), so it preserves any greedy or $\varepsilon$-greedy policy based on $Q$ values and identifiability. It is important

## Algorithm 9.4 Multi-agent dueling deep Q-learning for power allocation optimisation with experience replay

1:  Initialisation:
2:  **for** $i \in N$ **do**
3:     Randomly initialise the Q-network $Q(s, a; w)$
4:     Randomly initialise the target Q-network $\hat{Q}(s, a, \hat{w})$
5:     Initialise the replay memory $D$ with capacity $C$
6:  **end for**
7:  **while** Not convergence **do**
8:     **for** U2U $i, i \in N$ **do**
9:        **for** Iteration **do**
10:          Select the action $a_i^t$ using $\varepsilon$-greedy via $Q_i(s_i^t, a_i^t; w)$
11:          Measure SINR at the receiver according to (9.3)
12:          Update the reward $r_i^t$ according to (9.20)
13:          Observe the new state $s_i^{t+1}$
14:          Store transition $(s_i^t, a_i^t, r_i^t, s_i')$ in $D$
15:          **if** The replay memory D is full **then**
16:             Sample a uniform random mini-batch of $K$ transitions from $D$
17:             Combine the value function and advantage functions as in (9.28)
18:             Update parameters of neural networks from learning batches using stochastic gradient descent to minimise the loss:

$$\left[ r_i + \gamma Q(s', \max_{a' \in \mathcal{A}} \hat{Q}_i(s_i', a_i')) - Q(s_i, a_i) \right]^2 \qquad (9.29)$$

19:          **end if**
20:          Update the state $s_i^t = s_i^{t+1}$
21:       **end for**
22:    **end for**
23: **end while**

to note that (9.28) is viewed and implemented as part of the network but not as a separate algorithmic step. Moreover, estimations of $V(s)$ and $A(s, a)$ are computed automatically without extra supervision or algorithmic modifications. The details of the power allocation based on the dueling deep Q-learning algorithm are presented in Algorithm 9.4.

## 9.6 Simulation results

In this section, we evaluate the performance of the proposed methods on PC Intel(R) Core(TM) i7-8700 CPU @ 3.20Ghz. All algorithms are implemented using Tensorflow 1.13.1 [35]. The action-value function and its target networks are initialised

Table 9.1 Simulation parameters

| Parameters | Value |
| --- | --- |
| Bandwidth ($W$) | 1 MHz |
| Path-loss exponent | $\alpha_h = 3$ |
| $p_{max}$ | 23 dBm |
| $p_{min}$ | 0 dBm |
| Channel power gain at the reference | $\beta_0 = -30$ dB |
| Noise power density | $\eta = 0.5$ |
| SINR QoS constraint | $\gamma^* = 0$ dB |
| Learning rate | $\alpha = 0.01$ |
| Discount factor | $\gamma = 0.9$ |

with one hidden layer of 20 units. For initial values of weights and biases, we set to small random values according to a zero-mean Gaussian distribution with a standard deviation of 0.1. We use Adam optimiser [36] for training. The other simulation parameters are provided in Table 9.1.

## 9.6.1 Performance comparison

Figure 9.1 plots the performance results of four multi-agent DRL approaches with a different number of U2U pairs in coverage, $N = 2, 5, 7, 10$. Double deep Q-learning constantly performs better than other approaches in terms of the expected sum rate. Multi-agent classical Q-learning is less favourable due to its worst performance and high variance. At convergence, it seems that the performances of multi-agent double deep Q-learning and deep dueling deep Q-learning are comparable. These results suggest that using DRL and multi-agent learning significantly helps to find an optimal policy for non-cooperative energy-efficient power allocation in U2U communication.

## 9.6.2 Scalability analysis

We introduce another experiment to analyse how deep RL-based methods are able to scale to large domains. We fix the number of pairs $N = 10$ but vary the size of the action space. It is well known that the problem complexity increases exponentially with the size of the action space. Figure 9.2 compares the efficiency of the classical Q-learning and DRL approaches using a different size for the action space. The results show that the standard Q-learning approach is only suitable for a problem of small state and action space. On more complicated problems, the performance of Q-learning becomes very unstable. Meanwhile, the deep Q-learning, DDQL, and dueling DQL can still perform well and stabilise with larger action space $\mathscr{A}$.

## 9.6.3 Exploration/exploitation analysis

The balance between exploration and exploitation of DRL is a great challenge for use in practice. In this experiment, we measure the performance of DRL algorithms

*Non-cooperative EE power allocation game in U2U communication* 133

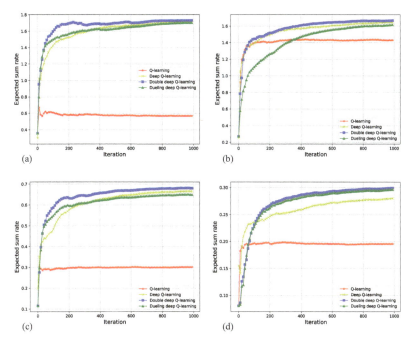

*Figure 9.1  Expected sum rate of the network with different number of U2U pairs: (a) $N = 10$, (b) $N = 7$, (c) $N = 5$, (d) $N = 2$, with the action space of size $|\mathscr{A}| = 23$ and $\varepsilon$-greedy $\varepsilon = 0.9$*

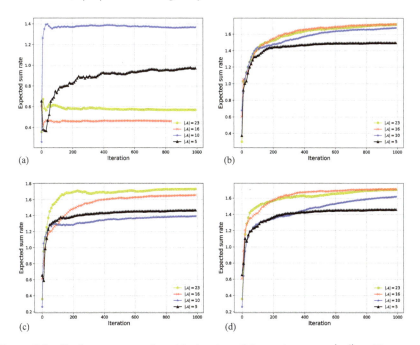

*Figure 9.2  Performance results with the size of the action space $|\mathscr{A}| = 23$, $|\mathscr{A}| = 16$, $|\mathscr{A}| = 10$, $|\mathscr{A}| = 5$: (a) Q-learning, (b) deep Q-learning, (c) double deep Q-learning, and (d) dueling deep Q-learning*

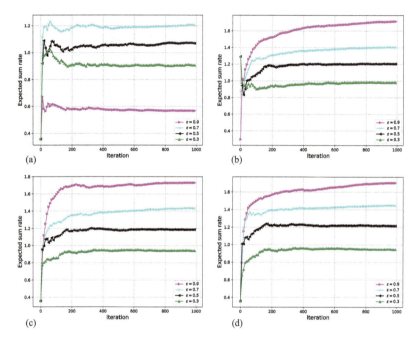

*Figure 9.3  Performance results with different exploration strategies ε-greedy: (a) Q-learning, (b) deep Q-learning, (c) double deep Q-learning, and (d) dueling deep Q-learning, on action space $|\mathcal{A}| = 23$ and number of U2U pairs $N = 10$*

Table 9.2  The running time of DRL in different environment scenarios

|  | $N = 2$ | $N = 5$ | $N = 7$ | $N = 10$ |
|---|---|---|---|---|
| $|\mathcal{A}| = 5$ | 0.39 ms | 1 ms | 1.38 ms | 1.95 ms |
| $|\mathcal{A}| = 10$ | 0.41 ms | 1.02 ms | 1.40 ms | 1.98 ms |
| $|\mathcal{A}| = 16$ | 0.42 ms | 1.02 ms | 1.41 ms | 2.02 ms |
| $|\mathcal{A}| = 23$ | 0.42 ms | 1.02 ms | 1.42 ms | 2.02 ms |

with different values of exploration rate $\varepsilon$. Figure 9.3 shows the comparison results for four proposed algorithms. The results suggest that all three DRL approaches are less sensitive to the choice of $\varepsilon$. That means when this parameter is set to 0.9, they perform quite stable.

### 9.6.4 Running time analysis

This experiment evaluates the running time of DRL during test time after training. Table 9.2 shows the average running time of DRL methods. Each U2U transmitter only needs a few milliseconds to select actions that have already been optimised offline. The selected actions are expected to maximise the performance of the network while satisfying the QoS constraints. The results suggest that our proposed methods are applicable for real-time optimisation and practical use.

## 9.7 Summary

This chapter presented non-cooperative DRL algorithms to optimise energy-efficient power allocation in U2U communication. In particular, it illustrated three different multi-agent DRL algorithms, showing how DRL can maximise the entire network's performance while guaranteeing the QoS constraints in real-time scenarios. In addition, the proposed approaches proved very promising regarding scalability to large domains, high performance, and low running time. Future research directions include investigating other DRL algorithms that can deal with continuous action problems and possible ways to improve multi-agent learning strategies for non-cooperative problems.

## References

[1] Nguyen KK, Duong TQ, Vien NA, *et al.* Non-cooperative energy efficient power allocation game in D2D communication: a multi-agent deep reinforcement learning approach. *IEEE Access*. 2019;7:100480–100490.

[2] Nguyen MT, Nguyen HM, Masaracchia A, *et al.* Stochastic-based power consumption analysis for data transmission in wireless sensor networks. *EAI Endorsed Trans Ind Netw Intell Syst*. 2019;6(19):1–11.

[3] Sutton RS, Barto AG, *et al. Introduction to Reinforcement Learning*. vol. 135. MIT Press Cambridge; 1998.

[4] Matarić MJ. Reinforcement learning in the multi-robot domain. In: *Robot Colonies*. Springer; 1997. p. 73–83.

[5] Steels L. Language as a complex adaptive system. In: *Int. Conf. Parallel Prob. Solv. Nature*. Springer; 2000. p. 17–26.

[6] Ghory I. Reinforcement learning in board games. *Tech Rep.*; 2004. p. 105.

[7] D. Silver, A. Huang, C. J. Maddison, A. Guez, L. Sifre, G. van den Driessche *et al.* "Mastering the game of Go with deep neural networks and tree search", *Nature*, vol. 529, no. 7587, pp. 484–489, Jan. 2016.

[8] LeCun Y, Bengio Y, and Hinton G. Deep learning. *Nature*. 2015;521(7553): 436–444.

[9] Parkhi OM, Vedaldi A, and Zisserman A. Deep face recognition. In: *bmvc*. vol. 1; 2015. p. 6.

[10] Krizhevsky A, Sutskever I, and Hinton GE. Imagenet classification with deep convolutional neural networks. In: *Adv. Neural Inf. Process. Syst.*; 2012. p. 1097–1105.

[11] Schroff F, Kalenichenko D, and Philbin J. Facenet: a unified embedding for face recognition and clustering. In: *Proc. IEEE Conf. Comp. Vision Pat. Recog.*; 2015. p. 815–823.

[12] Hu G, Yang Y, Yi D, et al. When face recognition meets with deep learning: an evaluation of convolutional neural networks for face recognition. In: *Proc. IEEE Int. Conf. Comp. Vision Workshop*; 2015. p. 142–150.

[13] Nguyen KK, Hoang DT, Niyato D, et al. Cyberattack detection in mobile cloud computing: a deep learning approach. In: *Proc. IEEE Wireless Commun. and Networking Conf.* 2018; 2018. p. 1–6.

[14] Mnih V, Kavukcuoglu K, Silver D, et al. *Playing atari with deep reinforcement learning*; 2013. Available from: https://arxiv.org/abs/1312.5602.

[15] Zhang L, Xiao M, Wu G, et al. Efficient scheduling and power allocation for D2D-assisted wireless caching networks. *IEEE Trans Commun*. 2016;64(6):2438–2452.

[16] Yin R, Zhong C, Yu G, et al. Joint spectrum and power allocation for D2D communications underlaying cellular networks. *IEEE Trans Veh Technol*. 2016 Apr;65(4):2182–2195.

[17] Jiang Y, Liu Q, Zheng F, et al. Energy-efficient joint resource allocation and power control for D2D communications. *IEEE Trans Veh Technol*. 2016;65(8):6119–6127.

[18] Nguyen MN, Nguyen LD, Duong TQ, et al. Real-time optimal resource allocation for embedded UAV communication systems. *IEEE Wireless Commun Lett*. 2019;8(1):225–228.

[19] Zhang P, Gao AY, and Theel O. Bandit learning with concurrent transmissions for energy-efficient flooding in sensor networks. *EAI Endorsed Trans Ind Netw Intell Syst*. 2018;4(13):1–14.

[20] Nguyen LD, Kortun A, and Duong TQ. An introduction of real-time embedded optimisation programming for UAV systems under disaster communication. *EAI Endorsed Trans Ind Netw Intell Syst*. 2018;5(17):1–8.

[21] Amiri R, Mehrpouyan H, Fridman L, et al. A machine learning approach for power allocation in HetNets considering QoS. In: *Proc. IEEE ICC 2018*; 2018. p. 1–7.

[22] Masadeh A, Wang Z, and Kamal AE. Reinforcement learning exploration algorithms for energy harvesting communications systems. In: *Proc. IEEE ICC 2018*; 2018. p. 1–6.

[23] Wang Y, Liu M, Yang J, et al. Data-driven deep learning for automatic modulation recognition in cognitive radios. *IEEE Trans Veh Technol*. 2019;68(4):4074–4077.

[24] Huang H, Xia W, Xiong J, et al. Unsupervised learning-based fast beamforming design for downlink MIMO. *IEEE Access*. 2019;7:7599–7605.

[25] Huang H, Song Y, Yang J, et al. Deep-learning-based millimeter-wave massive MIMO for hybrid precoding. *IEEE Trans Veh Technol*. 2019;68(3):3027–3032.

[26] Gui G, Huang H, Song Y, et al. Deep learning for an effective nonorthogonal multiple access scheme. *IEEE Trans Veh Technol*. 2018;67(9):8440–8450.

[27] Huang H, Guo S, Gui G, et al. Deep learning for physical-layer 5G wireless techniques: opportunities, challenges and solutions; 2019. Available from: arXivpreprintarXiv:1904.09673.

[28] Liu M, Song T, Hu J, *et al.* Deep learning-inspired message passing algorithm for efficient resource allocation in cognitive radio networks. *IEEE Trans Veh Technol.* 2019;68(1):641–653.
[29] Liu M, Song T, and Gui G. Deep cognitive perspective: Resource allocation for NOMA based heterogeneous IoT with imperfect SIC. *IEEE Internet Things J.* 2019;6(2):2885–2894.
[30] Sun Y, Peng M, and Mao S. Deep reinforcement learning based mode selection and resource management for green fog radio access networks. *IEEE Internet Things J.* 2019;6:1960–1971.
[31] Ye H and Li GY. Deep reinforcement learning for resource allocation in V2V communications. In: *Proc. IEEE ICC 2018*; 2018. p. 1–6.
[32] Bertsekas DP. *Dynamic Programming and Optimal Control.* vol. 1. Athena Scientific Belmont, MA; 1995.
[33] van Hasselt H, Guez A, and Silver D. Deep reinforcement learning with double Q-learning. In: *AAAI Conf Artif Intell.* 2016.
[34] Wang Z, Schaul T, Hessel M, *et al. Dueling network architectures for deep reinforcement learning*; 2015. Available from: https://arxiv.org/abs/1511.06581.
[35] Abadi M, Barham P, Chen J, Chen Z, Davis A, Dean J, Devin M, *et al.* "Tensorflow: A system for large-scale machine learning," in *Proc. 12th USENIX Sym. Opr. Syst. Design and Imp. (OSDI 16)*, Nov. 2016, pp. 265–283.
[36] Kingma DP and Ba JL. *Adam: A method for stochastic optimization*; 2014. Available from: https://arxiv.org/abs/1412.6980.

*Chapter 10*
# Real-time energy harvesting-aided scheduling in UAV-assisted D2D networks

Thanks to unmanned aerial vehicle (UAV) agility, UAV-assisted device-to-device (D2D) communications represent a communication paradigm that can be easily deployed within the next-generation 6G network. This type of network paradigm is expected to enhance user experience and performance by exploiting the direct D2D interaction supported by UAVs, especially at public events. However, the constant mobility of D2D ground users and the limited energy and flying time of UAVs represent impediments to the implementation of such network architecture for real-time applications. To tackle this issue, a novel model based on deep reinforcement learning is illustrated in this chapter. Such DRL-based model aims to find the optimal solution for energy harvesting time scheduling in UAV-assisted D2D communications. To make the system model more realistic, it is assumed that the UAV flies around a central point, the D2D users move continuously with a random walk model, and the channel state information encountered during each time slot is randomly time-variant. Our numerical results demonstrate that the proposed schemes outperform the existing solutions in terms of achievable energy efficiency and computational time, making it suitable for solving real-time resource allocation problems in UAV-assisted wireless networks.*

## 10.1  Introduction

Unmanned aerial vehicles (UAVs) have various wireless applications ranging from public safety, environmental monitoring, and enhanced network connectivity as a benefit of their nimble mobility features. UAVs are capable of assisting wireless networks in providing ubiquitous coverage, robust handovers, and flawless real-time multi-media communications. However, the performance of UAV-assisted networks is limited by the UAVs' energy-storage and the resultant flying time. Recent research has tackled some of the challenges in UAV-supported wireless communications [2–10]. Yet, most techniques rely on unrealistic simplifications and focus predominantly on data transmission. Hence, it is crucial to find solutions to the associated problems in realistic dynamic environments, as detailed below.

*This chapter has been published partly in [1].

Device-to-device (D2D) communications solutions have been designed for diverse applications, such as smart city operation [11] and video streaming [12], by exploiting direct D2D connections. UAV-supported D2D communications are eminently suitable for providing emergency notifications or simple text messages when deployed in inaccessible disaster zones or remote areas. But again, they potentially suffer from limited UAV flying time as well as energy constraints and often have strict computational deadlines in support of D2D communications. These stringent requirements call for powerful solutions in support of real-time resource allocation to enhance network performance while satisfying all the constraints. Several techniques have been proposed for solving the associated resource management problems, some of which achieve excellent performance, but they cannot satisfy the stringent time constraints of real-life applications.

Very recently, deep reinforcement learning (DRL)-based methods, which rely on a combination of reinforcement learning and neural networks, have demonstrated impressive results in resource allocation [13–16]. Upon interacting directly with the environment to learn by trial and error, the approaches based on DRL algorithms have demonstrated self-organising capability to adapt to a dynamic environment that exhibits rapidly fluctuating channel state information. Furthermore, by using neural networks for training, flexible and prompt decisions may be made according to the environment's state. Inspired by the success of DRL algorithms in solving resource allocation problems tackled in [13–16], we also rely on the broad family of DRL algorithm-based techniques to deal with the energy harvesting scheduling problem of UAV-assisted D2D communications. The UAV is considered to be an agent that interacts with the environment in order to find the optimal policy. After being trained, the agent embarks on the most appropriate action at each step by deciding to opt for either energy harvesting or information transmission, which will approximately maximise the network performance. We will demonstrate that the proposed models outperform the benchmarks in terms of the processing time imposed, requiring less than one millisecond to decide upon the most beneficial action for the next time step.

## *10.1.1 State of the art and challenges*

UAVs have been adopted for various applications such as geographic surveys, security control, agriculture, and goods delivery. For example, as seen in [17], Amazon provides a service that allows customers to opt for UAV delivery and receive packages within 30 minutes. In [18], a scouting UAV suitable for counter-terrorism was developed in order to discover weapons and hide-out locations, or to conduct a battle damage assessment. Moreover, UAVs are becoming practical resources for rescue teams and medical emergencies, as in payload delivery missions [19]. UAVs have also been used to enhance wireless network performance [4,6–8,20,21] or in disaster relief networks [4,20], public safety communications [22], and sensor data tracking systems [20,23]. For example, the authors of [8] proposed a cooperative interference cancellation scheme for multi-beam UAV communication, with the objective of maximising the uplink sum-rate, while suppressing the interference at the ground base stations (BSs). But again, UAVs also suffer from some stringent constraints owing to limitations in their power, flying time, and reliability. Thus, it is necessary to optimise their flight trajectory and resource allocation in order to enhance the network performance.

Sophisticated techniques have been developed to deal with resource allocation problems in UAV-assisted wireless networks. In [20], the authors proposed a system model using UAVs in the aftermath of natural disasters. A real-time optimisation method with low complexity was proposed to design the optimal path for gathering data in wireless sensor networks. In [23], a UAV was used to collect data from sensor nodes in Internet-of-Things (IoT)'s networks. By jointly optimising the sensor nodes's wake-up schedule and the UAV trajectory, the authors minimised energy consumption while reducing the data collection time. In [4], the authors provided network coverage to inaccessible disaster-stricken areas with the support of UAVs. The K-means clustering method, combined with a sophisticated power allocation algorithm, is proposed for the real-time support of users who send information about their positions and conditions to families and rescue teams during and after a disaster. However, there is still a paucity of solutions for realistic real-time scenarios.

To elaborate a little further, the associated resource allocation problems are complex owing to the dynamic positions of UAVs. In this context, the authors of [9] used the classic Lagrangian relaxation method to incorporate their constraints into the objective function and conceived resource optimisation for harvest-and-transmit protocols of UAV-assisted D2D communications. Explicitly, the D2D transmitters harvested energy from the UAV and then used the harvested energy for information transmission to the D2D receivers. In [7], a novel solution based on a logarithmic inequality was introduced for jointly optimising the power allocation and energy harvesting in UAV-aided D2D communications.

Both DL and RL have become popular for mitigating the violently fluctuating channel quality effects of realistic wireless networks. In [4], the authors proposed an unsupervised learning algorithm, namely the K-means algorithm, for clustering the remote users located in a disaster area into small groups and then a UAV was deployed to serve them separately. Indeed, numerous DRL algorithms have been proposed in the literature [13–16,24] for finding the optimal policy in the face of the near-instantaneously fluctuating propagation environment. The agents observe the environment's state and take action; then, trial-and-error-based learning is employed until the performance saturates. In realistic scenarios, it would take excessive time to find the optimal solution when relying on mathematical models. By contrast, DRL-based model-free schemes are potentially capable of finding solutions more promptly through neural networks following a training session. In [16], the authors proposed models based on both deep Q-learning and double deep Q-learning and dueling deep Q-learning algorithms to solve a multi-agent-aided power allocation problem in D2D communications. However, the action space in those algorithms has to be discrete, and human intervention may also be required. In [14], the authors improved the model by introducing a distributed deep deterministic policy gradient algorithm to solve the power control problem of D2D-based vehicle-to-vehicle communications. Inspired by the encouraging results of the above applications, we conceive bespoke DRL algorithms for optimising the energy harvesting time scheduling of UAV-enabled D2D communications.

### 10.1.2 Contributions

Within this chapter, the following contributions are provided:

- We conceive a system model capable of reflecting the dynamic position of UAVs and the unknown channel state information (CSI).
- We then formulate our energy harvesting time scheduling problem for UAV-assisted D2D communications, where the UAV is considered to be an agent in the game. The agent will observe the environment's state and then take the action of approximately choosing the specific time span $\tau$ that maximises the energy efficiency (EE) of the network. In each step of the DRL algorithm, the UAV chooses the most appropriate action to take according to the change in the environment and the CSI.
- We propose a novel deep deterministic policy gradient (DDPG) algorithm for solving the energy efficiency optimisation game for the UAV-supported D2D scenario considered. Our method outperforms conventional optimisation techniques in terms of its EE vs. complexity, hence resulting in a reduced processing time.
- We further improve our model with the aid of an efficient sampling technique by appropriately adapting the proximal policy optimisation (PPO) algorithms of [25]. More explicitly, we rely on the clipping surrogate objective and Kullbacka-Leibler (KL) divergence penalty [25] for formulating the objective function of the PPO algorithm. These techniques improve the speed of convergence and are robust in terms of adapting to the changes of the environment.

In the remainder of our chapter, Section 10.2 formulates the system model of UAV-enabled D2D communications. The background of DRL algorithms is introduced in Section 10.3. In Section 10.4, we solve our energy harvesting time scheduling problem by using the DDPG algorithm for the continuous action space of D2D communications supported by a UAV. To improve training and sampling, in Section 10.5, we propose a model based on the PPO algorithm to constrain the size of policy updating at each iteration. Our numerical results are presented in Section 10.6 for characterising DRL algorithm-based UAV-assisted wireless networks. Finally, we conclude the chapter and propose some future research directions in Section 10.7.

## 10.2 System model and problem formulation

The considered system includes $N$ D2D pairs and one UAV equipped with a single-antenna, as seen in Figure 10.1. Each D2D pair consists of a single-antenna D2D transmitter (D2D-Tx) and a single-antenna D2D receiver (D2D-Rx). The $N$ D2D pairs are randomly distributed within the UAV's coverage area. In each time step, the D2D pairs are moving continuously following the random walk model with the velocity $v$. The UAV is moving randomly in a zone around a point in the centre

*EH-aided scheduling in UAV-assisted D2D networks* 143

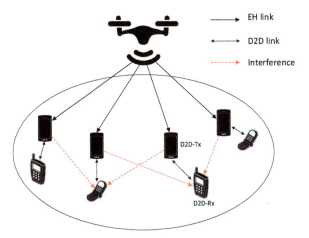

*Figure 10.1 System model of D2D communications supported by a UAV*

of coverage due to the effect of wind. We adopt the harvest-then-transmit protocol presented in [26]. The energy harvesting and information transmission take place in their dedicated phases. In the first phase, the D2D-Tx harvests energy from the UAV during time span $\tau \mathbb{T}$ with $0 \leq \tau \leq 1$. Then in the remaining phase spanning $(1 - \tau)\mathbb{T}$ the information transmission between D2D-Tx and D2D-Rx takes place. For convenience, we assume that the block time is normalised to 1, $\mathbb{T} = 1$.

We assume that the central point in the 3D place of the UAV is $(x_0, y_0, H_0)$, where $H_0$ is the altitude of the UAV's antenna. In a real-life application, the UAV is affected by environmental factors, such as wind and rain, which is the reason for its random movement around the position $(x_{UAV}, y_{UAV}, H_{UAV})$. The locations of the $i$th D2D-Tx and the $j$th D2D-Rx are $(x_i^{Tx}, y_i^{Tx})$ and $(x_i^{Rx}, y_i^{Rx})$ with $i, j = 1, \ldots, N$, respectively. The distance between the UAV and the $i$th D2D-Tx is given by:

$$D_i = \sqrt{d_i^2 + H_{UAV}^2}, \tag{10.1}$$

where $d_i = \sqrt{(x_{UAV} - x_i^{Tx})^2 + (y_{UAV} - y_i^{Tx})^2}$ is the Euclidean distance between the UAV and the $i$th D2D-Tx. The air-to-ground channel between the UAV and the D2D-Tx is subjected to blockage effects of buildings. The probability of having a line-of-sight (LoS) connection between the UAV and the $i$th D2D user is given by [27]:

$$P_i^{LoS} = \frac{1}{1 + a \exp\left[-b(\Theta_i - a)\right]}, \tag{10.2}$$

where $a$ and $b$ are constants that depend on the environment. The elevation angle $\theta$ is defined as:

$$\Theta_i = \frac{180}{\pi} \sin^{-1}\left(\frac{H_{UAV}}{D_i}\right). \tag{10.3}$$

The probability of having a non-line-of-sight (NLoS) link is $P_i^{NLoS} = 1 - P_i^{LoS}$. Thus, the channel gain between the UAV and the $i$th D2D user characterised by the LoS and NLoS is given by:

$$h_i = P_i^{LoS} \times d_i^{-\alpha_E} + P_i^{NLoS} \times \omega d_i^{-\alpha_E}, \tag{10.4}$$

where $\omega$ and $\alpha_E$ are the NLOS connection factor and the path-loss exponent between the UAV and the user link, respectively.

Furthermore, we define $P_0$ and $g_i$ as the maximum total transmit power of the UAV and the link's power gain between the UAV and the $i$th D2D-Tx, respectively. The energy harvested at the $i$th D2D-Tx over the time span $\tau$ is given by:

$$\mathcal{E} = \tau \eta P_0 g_i, \tag{10.5}$$

where $0 < \eta < 1$ is the energy harvesting efficiency. We denote the information transmission power usage between D2D-Tx and D2D-Rx of the $i$th D2D pair by $p_i$. Thus, the total power usage available for information transmission in the entire network is given by:

$$\Phi = \sum_i^N (1 - \tau) p_i, i \in N. \tag{10.6}$$

We define the channel power gain at the reference distance by $\beta_{D2D}$, the small-scale fading channel power gain (an exponentially distributed random variable) by $f_i^2$, and the path-loss exponent by $\alpha$. The channel's power gain between the $i$th D2D-Tx and the $j$th D2D-Rx is defined as:

$$h_{ij} = \beta_{D2D} f_i^2 d_{ij}^{-\alpha}, \tag{10.7}$$

where $d_{ij} = \sqrt{(x_i^{Tx} - x_j^{Rx})^2 + (y_i^{Tx} - y_j^{Rx})^2}$ is the Euclidean distance between $i$th D2D-Tx and $j$th D2D-Rx.

The signal-to-interference-plus-noise (SINR) ratio at the $i$th D2D user's receiver is as follows:

$$\gamma_i = \frac{p_i h_{ii}}{\sum_{j \in N}^{j \neq i} p_j h_{ji} + \sigma^2}, \tag{10.8}$$

where $h_{ji}$ is the channel gain between the $j$th D2D-Tx and the $i$th D2D-Rx, while $\sigma^2$ is the AWGN power.

The information throughput at the $i$th D2D pair is given by:

$$R_i(\tau, p_i) = (1 - \tau) W \log_2 (1 + \gamma_i), \tag{10.9}$$

where $W$ is the bandwidth. The D2D link's communication constraint is formulated as follows:

$$R_i(\tau, p_i) \geq r_{min}, \forall i \in N, \tag{10.10}$$

where the threshold $r_{min}$ represents the quality-of-service (QoS) constraint. The power total consumption during the energy harvesting phase between the UAV and the $i$th D2D user, as well as of the information transmission phase between the $i$th D2D-Tx and the $i$th D2D-Rx, is formulated as:

$$\rho(\tau, p) = \sum_{i=1}^{N} (1-\tau)p_i + \tau\eta P_0 + P_{cir}, \quad (10.11)$$

where we have $p = [p_i]|_{i=1}^{N}$ and $P_{cir}$ is the total circuit power dissipation at the UAV and the D2D users.

In harvesting scheduling optimisation, we assume that each D2D-Tx uses the maximum energy harvested from the UAV for transmitting information, yielding:

$$(1-\tau)p_i = \tau\eta P_0 g_i. \quad (10.12)$$

Our objective is to maximise the EE defined as:

$$\chi = \frac{\sum_{i=1}^{N} R_i(\tau, p_i)}{\rho(\tau, p)}, \quad (10.13)$$

Thus, we formulate the EE optimisation problem as:

$$\max_{\tau} \frac{\sum_{i=1}^{N}(1-\tau)W\log_2(1+\gamma_i)}{\sum_{i=1}^{N}(1-\tau)p_i + \tau\eta P_0 + P_{cir}}$$
$$\text{s.t. } 0 < \tau < 1 \quad (10.14)$$
$$R_i(\tau, p_i) \geq r_{min}, \forall i \in N,$$

We proceed by setting up our EE game as a Markov decision process (MDP) [28], defined by the five-tuple $\langle \mathscr{S}, \mathscr{A}, \mathscr{P}, \mathscr{R}, \zeta \rangle$, where $\mathscr{S}$ is the state space, $\mathscr{A}$ is the action space, $\mathscr{P}: \mathscr{S} \times \mathscr{A} \times \mathscr{S} \to \mathbb{R}$ is the state transition probability, $\mathscr{R}: \mathscr{S} \to \mathbb{R}$ is the reward function, and $\zeta \in (0, 1)$ is the discount factor. We define the policy mapping as a distribution $\pi: \mathscr{S} \times \mathscr{A} \to [0, 1]$. Then, the game is formulated as follows:

- *Agent*: The UAV is an agent. The agent observes the state and takes action to interact with the environment to find the optimal policy.
- *State space*: The state space is defined as a cooperative state of all nodes in the network as:

$$\mathscr{S} = \{\mathscr{I}_1, \ldots, \mathscr{I}_i, \ldots \mathscr{I}_N\}, \quad (10.15)$$

where $\mathscr{I}_i$ indicates whether the $i$th D2D pair satisfies the SINR constraints:

$$\mathscr{I}_i = \begin{cases} 1 & \text{for } R_i(\tau, p_i) \geq r_{min} \\ 0 & \text{for otherwise.} \end{cases} \quad (10.16)$$

- *Action space*: The agent at state $s$ selects an action $a$ from the legitimate action space to obtain the reward $r$,

$$\mathscr{A} = \{\tau\}, 0 < \tau < 1. \quad (10.17)$$

- *Reward function*: At each time step $t$, the agent will take action $a$ following the policy $\pi$ to maximise the reward $r$ of the network. The reward function is a joint function of all D2D pairs formulated as:

$$\mathscr{R} = \frac{\sum_{i=1}^{N}(1-\tau)W\log_2(1+\gamma_i)}{\sum_{i=1}^{N}(1-\tau)p_i + \tau\eta P_0 + P_{cir}}. \quad (10.18)$$

## 10.3 Preliminaries

The DRL algorithms rely either on a value function-based or policy gradient-based model. In this section, we briefly present the concepts and mathematical formulations of both the value function and the policy gradient-based models.

### 10.3.1 Value function

The value function at state $s$ is the expected reward, while following policy $\pi$:

$$V^{\pi}(s) = \mathbb{E}\left[\sum_{t\geq 0} \zeta r^t | s_0 = s, \pi\right], \tag{10.19}$$

where the expectation $\mathbb{E}[\,.\,]$ denotes the empirical average over a sample batch.

The action-value function is the expected reward obtained after taking action $a$ at state $s$ under the policy $\pi$, which is expressed as:

$$Q^{\pi}(s, a) = \mathbb{E}\left[\sum_{t\geq 0} \zeta r^t | s_0 = s, a_0 = a, \pi\right]. \tag{10.20}$$

The optimal performance obtained at state $s$ when taking action $a$ is associated with the maximum expected reward defined as:

$$Q^*(s, a) = Q^{\pi^*}(s, a) = \max_{\pi} \mathbb{E}\left[\sum_{t\geq 0} \zeta r^t | s_0 = s, a_0 = a, \pi\right], \tag{10.21}$$

where $\pi^*$ is the optimal policy.

We can reach the optimal performance by finding the optimal policy $\pi^*$ that follows the Bellman equation [29]:

$$Q^*(s, a) = \mathbb{E}\left[r + \zeta \max_{a'} Q^*(s', a') | s, a\right]. \tag{10.22}$$

The deep Q-learning algorithm of [30] has recently gained substantial attention since it is eminently suitable for estimating the action-value function. To estimate the action-value function, we use a function approximator $Q(s, a; \theta) \approx Q^*(s, a)$ where $\theta$ is the parameter of the neural network. Our objective is deep Q-learning is that of minimising the loss $L_i(\theta_i)$ at each iteration $i$ as follows:

$$L_i(\theta_i) = \mathbb{E}\left[(y_i - Q(s, a; \theta_i))^2\right], \tag{10.23}$$

where $\hat{Q}(s', a'; \theta_{\text{targ}})$ is the target network with parameter $\theta_{\text{targ}}$ and the target value $y_i$ for iteration $i$ is defined as:

$$y_i = \mathbb{E}\left[r + \zeta \max_{a'} \hat{Q}(s', a'; \theta_{\text{targ}}) | s, a\right]. \tag{10.24}$$

The gradient update is written as:

$$\nabla_{\theta_i} L_i(\theta_i) = \mathbb{E}\left[r + \zeta \max_{a'} \hat{Q}(s', a'; \theta_{\text{targ}}) - Q(s, a; \theta_i) \nabla_{\theta_i} Q(s, a; \theta_i)\right]. \quad (10.25)$$

### 10.3.2 Policy search

As for the policy search-based method, we can directly search for an optimal policy $\pi^*$ for the agents to reach the best performance in terms of maximising the reward value of:

$$J(\theta) = \mathbb{E}\left[\sum_{t \geq 0} \zeta r^t | \pi_0\right]. \quad (10.26)$$

The optimal policy parameters can be formulated as:

$$\phi^* = \arg\max J(\phi). \quad (10.27)$$

Mathematically, the average value can be written as:

$$J(\phi) = \int^\kappa R(\kappa) p(\kappa, \phi) d\kappa, \quad (10.28)$$

where $\kappa$ is represented by the trajectory $\{s^0, a^0, s^1, a^1, \ldots, s^{T-1}, a^T, s^T\}$, while $p(\kappa; \phi)$ is a trajectory distribution given by:

$$p(\kappa; \phi) = p(s^0) \prod_{t=0}^{T-1} p(s^{t+1} | s^t, a^t) \pi(a^t | s^t; \phi), \quad (10.29)$$

where $\phi$ denotes the parameters of the policy $\pi$. Upon differentiating the expected reward, we have [31]:

$$\nabla_\phi J(\phi) = \int^\kappa R(\kappa) \nabla_\phi p(\kappa; \phi) d\kappa$$
$$= \int^\kappa R(\kappa) \nabla_\phi \log p(\kappa; \phi) p(\kappa; \phi) d\kappa \quad (10.30)$$
$$= \mathbb{E}\left[R(\kappa) \nabla_\phi \log p(\kappa; \phi)\right].$$

Thus, we can estimate the gradient of the reward function by:

$$\nabla_\phi J(\phi) \approx \sum_{t=0}^{T-1} R(\kappa) \nabla_\phi \log \pi(a^t | s^t; \phi). \quad (10.31)$$

The parameter $\phi$ corresponding to the policy $\pi$ can be updated by using the stochastic gradient descent algorithm as:

$$\phi \leftarrow \phi + \alpha \nabla_\phi J(\phi), \quad (10.32)$$

where $\alpha \in [0, 1]$ is the learning rate.

Several algorithms have been developed in the literature based on policy search, such as natural policy gradient methods [32] and vanilla policy gradient methods [33].

## 10.4 Energy harvesting time scheduling in UAV-powered D2D communications: a deep deterministic policy gradient approach

This section proposes a deep deterministic policy gradient algorithm (DDPG) [34] for energy harvesting time scheduling in UAV-powered D2D communications. The DDPG algorithm is a hybrid model of the value function and policy search methods. By exploiting the benefits of both models, the DDPG algorithm improves the convergence speed of the optimisation to be suitable even for large-scale action spaces.

The DDPG algorithm consists of two fundamental elements: the actor and critic functions. The actor function $\mu(s; \theta_\mu)$ maps the states to a specific action according to the current policy, while the critic function $Q(s, a)$ is learned as in Q-learning for qualifying the action taken. The pair of techniques that we advocate in the DRL algorithm are as follows:

- *Experience replay buffer*: We use a replay memory pool $\mathscr{D}$ for storing the transitions $(s^t, a^t, r^t, s^{t+1})$ inferred from the environment under the exploration policy. A mini-batch $K$ of samples stored in the replay buffer $\mathscr{D}$ will be randomly taken for training the actor and critic network. The buffer $\mathscr{D}$ is also set to a finite size. The oldest transitions are discarded for updating samples space, hence the buffer is always up-to-date.
- *Target network*: One of the challenges during the training step is the unstable nature of the network if we use a shifting set of $Q$ values for calculating the target value. We use the target network to estimate the target values to overcome this challenge. Here particularly, in the DDPG algorithm, we use the target actor network and the target critic network, $\mu'(s; \theta_{\mu'})$ and $Q'(s, a; \theta_{q'})$, respectively.

We create a mini-batch of $K$ transitions $(s^k, a^k, r^k, s^{k+1})$ from the buffer $\mathscr{D}$ by random sampling for training. The critic network parameters are updated to minimising the loss function:

$$L = \frac{1}{K} \sum_{k}^{K} \left(y^k - Q(s^k, a^k; \theta_q)\right)^2, \tag{10.33}$$

where we have:

$$y^k = r^k(s^k, a^k) + \zeta Q'(s^{k+1}, a^{k+1}; \theta_{q'})|_{a^{k+1} = \mu'(s^{k+1}; \theta_{\mu'})}. \tag{10.34}$$

The actor policy is updated using the sampled policy gradient as follows:

$$\nabla_{\theta_\mu} J \approx \frac{1}{K} \sum_{k}^{K} \nabla_{a^k} Q(s^k, a^k; \theta_q)|_{a^k = \mu(s^k)} \nabla_{\theta_\mu} \mu(s^k; \theta_\mu). \tag{10.35}$$

The parameters $\theta_q$ and $\theta_{\mu'}$ of the target actor network and the target critic network are then updated by using soft target updates associated with $\varkappa \ll 1$ as illustrated below:

$$\theta_{q'} \leftarrow \varkappa\theta_q + (1 - \varkappa)\theta_{q'}, \tag{10.36}$$

$$\theta_{\mu'} \leftarrow \varkappa\theta_\mu + (1 - \varkappa)\theta_{\mu'}. \tag{10.37}$$

It makes the target values be constrained to change significantly more slowly, which allows the $Q$ function approach to supervised learning more closely. However, the price is that this may slow down the training due to the delayed value estimators propagation in the target networks $\mu'$ and $Q'$. We have to find a good exploration policy in a continuous action space to attain better convergence. Thus, we add a noise process of $\mathcal{N}(0, 1)$ associated with a small constant $\psi$ to our actor policy, which is formulated as [34]:

$$\mu'(s^t) = \mu(s^t; \theta_\mu^t) + \psi \mathcal{N}(0, 1). \tag{10.38}$$

The details of our DDPG algorithm-based technique of solving the energy harvesting time scheduling in UAV-assisted D2D communications is described in Algorithm 10.1 where $M$ and $T$ are the number of maximum episodes and time step per episode, respectively.

---

**Algorithm 10.1** Deep deterministic policy gradient algorithm for energy harvesting time scheduling in UAV-assisted D2D communications

---
1: Initialise the critic network $Q(s, a; \theta_q)$ and the actor network $\mu(s; \theta_\mu)$ with random parameter $\theta_q$ and $\theta_\mu$, respectively
2: Initialise the target critic networks $Q'$ and the target actor network $\mu'$ with parameter $\theta_{q'} \leftarrow \theta_q, \theta_{\mu'} \leftarrow \theta_\mu$, respectively
3: Initialise the replay memory pool $\mathcal{D}$
4: **for** episode = $1, \ldots, M$ **do**
5:     Initialise a random process $\mathcal{N}$ for the action exploration
6:     Receive initial observation state $s^0$
7:     **for** iteration = $1, \ldots, T$ **do**
8:         Obtain the action $a^t$ at state $s^t$ according to the current policy and the exploration noise
9:         Measure the achieved SINR according to (10.8)
10:         Update the reward $r^t$ according to (10.18)
11:         Observe the new state $s^{t+1}$
12:         Store transition $(s^t, a^t, r^t, s^{t+1})$ into the replay buffer $\mathcal{D}$
13:         Sample randomly a mini-batch of $K$ transitions $(s^k, a^k, r^k, s^{k+1})$ from $\mathcal{D}$
14:         Update the critic by minimising the loss as in (10.33)
15:         Update the actor policy using the sampled policy gradient as in (10.35)
16:         Update the target networks as in (10.36) and (10.37)
17:         Update the state $s_i^t = s_i^{t+1}$
18:     **end for**
19: **end for**

## 10.5 Efficient learning with proximal policy optimisation algorithms to solve the energy harvesting time scheduling problem in D2D communications assisted by UAV

In this section, we propose a novel model based on the PPO algorithm relying on an efficient sampling technique for solving the energy harvesting time scheduling game in UAV-assisted D2D communications. The PPO algorithm allows the policy to carry out the most significant possible improvement step using the current data without over-estimation, which might degrade the performance.

### 10.5.1 Clipping surrogate method

Let $p_\theta^t$ denote the probability ratio $p_\theta^t = \frac{\pi(s,a;\theta)}{\pi(s,a;\theta_{old})}$. Then our main objective is to maximise $\mathscr{L}$ as follows:

$$\mathscr{L}(s,a;\theta) = \mathbb{E}\left[\frac{\pi(s,a;\theta)}{\pi(s,a;\theta_{old})} A^\pi(s,a)\right] \quad (10.39)$$
$$= \mathbb{E}\left[p_\theta^t A^\pi(s,a)\right],$$

where $A^\pi(s,a) = Q^\pi(s,a) - V^\pi(s)$ is an estimator of the advantage function defined in [35]. Again, we create a mini-batch $K$ and then use the classic stochastic policy gradient descent to train our neural networks. The policy parameter is updated via:

$$\theta^{k+1} = \arg\max \mathbb{E}\left[\mathscr{L}(s,a;\theta^k)\right]. \quad (10.40)$$

Without improving any constraints, the maximisation of $\mathscr{L}(s,a;\theta)$ may lead to an incentive for the policy to move the probability $p_\theta^t$ away from 1. Thus, we opt for a suitable clipping technique for modifying the objective of (10.40) to [25]:

$$\mathscr{L}^{CLIP}(s,a;\theta) = \mathbb{E}\left[\min(p_\theta^t A^\pi(s,a), \; \text{clip}(p_\theta^t, 1-\varepsilon, 1+\varepsilon) A^\pi(s,a))\right], \quad (10.41)$$

where $\varepsilon$ is a small hyper-parameter. We use the function $\text{clip}(p_\theta^t, 1-\varepsilon, 1+\varepsilon)$ for limiting the probability ratio to avoid the excessive modification of $p_\theta^t$ outside the interval $[1-\varepsilon, 1+\varepsilon]$. In this chapter, we use an estimate of the advantage function $A^\pi(s,a)$ formulated as [33]:

$$A^\pi(s,a) = r^t + \zeta V^\pi(s^{t+1}) - V^\pi(s^t). \quad (10.42)$$

### 10.5.2 Kullback–Leibler divergence penalty

Instead of using clipping surrogate objective as in Section 10.5.1, we can also use the KL divergence penalty [25] based technique, where the parameters are updated by optimising the KL penalty objective [25]:

$$\mathscr{L}^{KL}(s,a;\theta) = \mathbb{E}\left[\frac{\pi(a|s;\phi)}{\pi(a|s;\phi_{old})} A(s,a) - \varphi KL[\pi(\cdot|s;\phi_{old}), \pi(\cdot|s;\phi)]\right]. \quad (10.43)$$

**Algorithm 10.2** Our proposed method based on the PPO algorithm for energy harvesting time scheduling in UAV-assisted D2D communications

1: Initialise the policy parameter $\theta_\pi$
2: Initialise the penalty method parameters
3: **for** episode = 1, ..., M **do**
4:     Receive initial observation state $s^0$
5:     **for** iteration = 1, ..., T **do**
6:         Obtain the action $a^t$ at state $s^t$ according to the current policy
7:         Measure the achieved SINR according to (10.8)
8:         Update the reward $r^t$ according to (10.18)
9:         Observe the new state $s^{t+1}$
10:        Update the state $s_i^t = s_i^{t+1}$
11:        Collect a set of partial trajectories with $K$ transitions
12:        Estimate the advantage function according to (10.42)
13:     **end for**
14:     Update the policy parameters using stochastic gradient descent with mini-batch $K$

$$\theta^{k+1} = \arg\max \frac{1}{K} \sum_{}^{K} \mathscr{L}(s, a; \theta^k) \quad (10.44)$$

15: **end for**

Then, we compute $d = \mathbb{E}\left[KL[\pi(\,\cdot\,|s; \phi_{old}), \pi(\,\cdot\,|s; \phi)]\right]$ based on the target value $d_{\text{targ}}$ of KL divergence [25]:

- if $d < d_{\text{targ}}/1.5$, $\varphi \leftarrow \varphi/2$,
- if $d < d_{\text{targ}} \times 1.5$, $\varphi \leftarrow \varphi \times 2$.

The parameter $\varphi$ is promptly updated in the next episode. The details of the PPO-based algorithm for solving the energy harvesting time scheduling in UAV-assisted D2D communications are presented in Algorithm 10.2.

## 10.6 Simulation results

In this section, we illustrate the efficiency of our DRL algorithms over the conventional approaches. All the algorithms are implemented using Tensorflow 1.13.1 [36] and the Adam optimisation algorithm [37] for training the neural networks. The algorithm in [7] is implemented using Python and CVXPY library [38] for convex optimisation. All the other simulation parameters are provided in Table 10.1.

### 10.6.1 Performance comparison

Figure 10.2 characterises our DRL algorithms when the number of D2D pairs is set to $N = 30$, in comparison to the optimal harvesting time optimisation (OHT) solution of

*Table 10.1  Simulation parameters*

| Parameters | Value |
| --- | --- |
| Bandwidth ($W$) | 1 MHz |
| UAV transmission power | 5 W |
| The central point of UAV | $(x_0, y_0, H_0) = (0, 0, 200)$ |
| Max moving distance of users per time step | 1 m |
| Max distance of UAV from central point | 1 m |
| Path-loss parameter | $\alpha_h = 3$ |
| D2D-Tx and D2D-Rx max distance | 50 m |
| Environment parameter | $a = 11.95, b = 0.136$ |
| Channel power gain | $\beta = -30$ dB |
| EH efficiency | $\eta = 0.5$ |
| The NLOS connection factor | $\omega = 20$ dB |
| Non-transmit power of UAV and D2D | 4 W |
| Clipping parameter | $\varepsilon = 0.2$ |
| Discounting factor | $\zeta = 0.9$ |
| Max number of D2D pairs | 30 |
| Initial batch size | $K = 32$ |
| Number of units per layer | 100 |

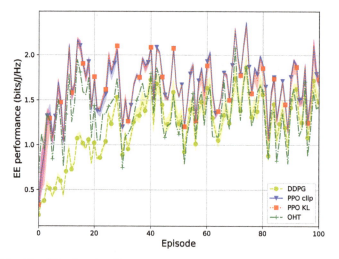

*Figure 10.2  The EE of optimal harvesting time using our DDPG, PPO algorithm, and the OHT optimisation [7] when the number of D2D pairs $N = 30$*

[7]. We use two hidden layers for the DDPG algorithm associated with 100 nodes per layer, while we use a single hidden layer associated with 100 nodes per layer for the PPO algorithm. This is because the DDPG algorithm has an off-policy nature, while the PPO algorithm is of an on-policy nature. The position of the UAV is changing over time, hence the channel state information also fluctuates dynamically in every time step. Both algorithms approach the optimal performance after about 50 episodes.

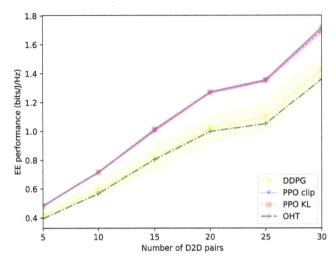

*Figure 10.3 The performance results of optimal harvesting time scheduling with different numbers of D2D pairs N*

The results based on our DDPG and PPO algorithm combination are better than those of the ones using OHT optimisation. Moreover, as can be observed in Figure 10.2, the methods based on both the PPO algorithm using the above-mentioned clip surrogate and KL divergence penalty achieve both similar EE and convergence speeds. The convergence speed of the scheme using the PPO algorithm is substantially faster than that of the ones using the DDPG algorithm.

In Figure 10.3, we present the performance of our DRL algorithm-based methods and the OHT optimisation-based method [7] while considering different numbers of D2D pairs. We average the performance of over 200 episodes. The average performance of the PPO algorithm-based method is higher than that of the DDPG algorithms. Furthermore, the EE of our proposed solutions is better than that of OHT optimisation, regardless of the number of D2D pairs within the UAV's coverage area.

The performance for different values of connection constraints of the DDPG and PPO algorithms is presented in Figure 10.4. The EE of methods based on the DDPG and PPO algorithm are similar when $r_{min}$ is in the range of 0.2 to 1.0. The results suggest that the PPO algorithm is indeed flexible and robust in the scenarios considered.

## 10.6.2 Parameter analysis

In Figure 10.5, we present the EE of our method based on the PPO algorithm for different batch sizes, $K$. Upon increasing the batch size $K$, the convergence speed is reduced. This is because when we take a smaller batch size of samples to train the networks in the PPO algorithm, the policy parameters are updated more frequently. Thus, we can approach the optimal performance faster. However, if we take the training time into consideration, the smaller batch size requires more time to train the

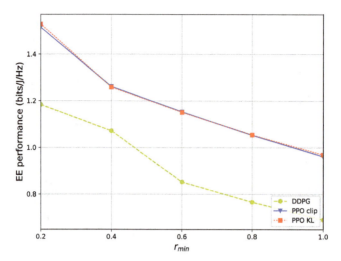

*Figure 10.4 The performance results of optimal harvesting time scheduling with different QoS constraints*

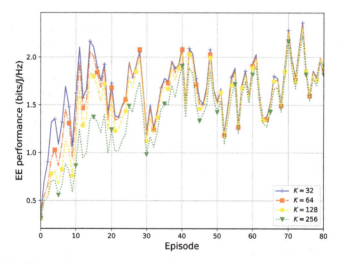

*Figure 10.5 The performance of optimal harvesting time using our PPO algorithm relying on the clipping objective technique while considering different values of batch size K*

neural networks in order to collect enough samples. In this study, we opted for the batch size of $K = 32$ for the implementation of the DDPG and PPO algorithms.

In Figure 10.6, we consider the difference in EE between the method using the DDPG algorithm in conjunction with various exploration parameters, $\psi$. Choosing the appropriate value of exploration is one of the challenges in designing the DDPG algorithm. If we choose the value $\psi$ to be too small in Figure 10.6, our algorithm

*EH-aided scheduling in UAV-assisted D2D networks* 155

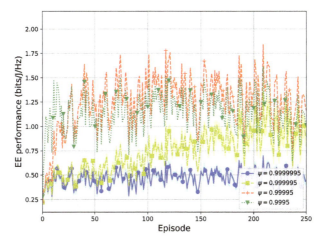

*Figure 10.6 The performance of optimal harvesting time using our DDPG algorithm with different values of exploration parameter $\psi$*

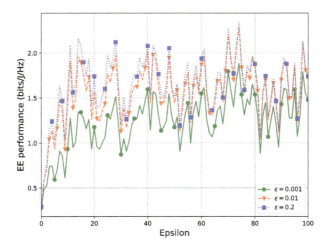

*Figure 10.7 The performance of optimal harvesting time by using our PPO algorithm relying on the clipping objective technique with different values of $\varepsilon$*

will be stuck at a local optimum because the DRL algorithm is a trial-and-error-based method. Hence, the agents cannot reach the optimal policy if we do not allow the agent to try all the possible circumstances. By contrast, if we choose an excessive exploration parameter, $\psi$, the convergence speed will be affected because the agents may bounce around the optimal value to explore more hitherto unexplored information. This reduces the convergence speed. As a compromise, we opted for the exploration ratio of $\psi = 0.99995$ for the DDPG algorithm.

Figure 10.7 presents the performance of UAV-assisted D2D communications upon using the PPO algorithm relying on the clipping surrogate method while

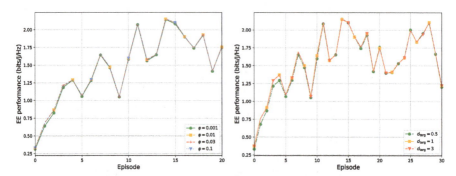

*Figure 10.8 The performance of optimal harvesting time by using our PPO algorithm relying on the KL divergence penalty technique*

considering different clipping thresholds, $\varepsilon$. The results show that we can achieve the best performance with a threshold of $\varepsilon = 0.2$. Meanwhile, Figure 10.8 illustrates the EE of the PPO algorithm using the KL penalty divergence in our optimal harvesting time scheduling problem of UAV-assisted D2D communications. In Figure 10.8a, the initial value of $\varphi$ is not critical in the PPO algorithm using the KL penalty method because $\varphi$ quickly adjusts. Thus, the results are similar for different values of $\varphi$. As a further result, Figure 10.8b shows the performance when we employ the method based on the PPO algorithm using the KL method with different values of $d_{\text{targ}}$. The results suggest that we should choose the value of $d_{\text{targ}}$ to be moderate for rapid convergence.

### 10.6.3 Computational complexity

We compare the computational complexity of our DDPG algorithm and PPO algorithm in the training phase for the energy harvesting time scheduling problem of D2D communications supported by a UAV. With the DDPG algorithm, the computational complexity is $O(MTK(Nn_{l1} + n_{l1}n_{l2} + \cdots))$ with $n_{li}$ is the number of nodes at layer $i$ in a neural network. On the one hand, the complexity is $O(MT + MK(Nn_{l1} + n_{l1}n_{l2} + \cdots))$ with the PPO algorithm. Furthermore, we compare the time processing of our neural networks in the testing phase and the OHT algorithm [7]. After training, simple calculations are required to predict the proper action in each time step. The computational complexity of trained networks is $O(Nn_{l1} + n_{l1}n_{l2} + \cdots)$. Specifically, our proposed DDPG algorithm solved the problem in only 0.229 ms, 0.259 ms, and 0.255 ms with the number of D2D pairs at 5, 15, and 30, respectively. Meanwhile, the OHT optimisation takes 54.1 ms, 115.5 ms, and 170.3 ms. Thus, our DDPG algorithm and PPO algorithm outperform mathematical models in terms of robustness, EE, and complexity (see Table 10.2).

## 10.7 Summary

This chapter illustrated efficient DRL algorithms to schedule the energy harvesting time in UAV-assisted D2D communications. It has been illustrated how such

Table 10.2 *The processing time while considering the varied number of D2D pairs*

|        | DDPG     | PPO clip | PPO KL   | OHT [7]  |
|--------|----------|----------|----------|----------|
| $N=5$  | 0.229 ms | 0.259 ms | 0.255 ms | 54.1 ms  |
| $N=15$ | 0.246 ms | 0.277 ms | 0.281 ms | 115.5 ms |
| $N=30$ | 0.260 ms | 0.276 ms | 0.286 ms | 170.3 ms |

approaches outperform benchmarks in EE and complexity, suggesting that they represent a promising approach for future real-time applications under the limitation of energy storage and flying time-constrained UAVs. Future research directions include investigating approaches for solving more complicated problems like joint optimisation of power allocation and trajectory planning and scenarios where multiple UAVs are included.

# References

[1] Nguyen KK, Vien NA, Nguyen LD, *et al*. Real-Time Energy Harvesting Aided Scheduling in UAV-Assisted D2D Networks Relying on Deep Reinforcement Learning. *IEEE Access*. 2021;9:3638–3648.

[2] Duan R, Wang J, Jiang C, *et al*. The Transmit-Energy vs Computation-Delay Trade-Off in Gateway-Selection for Heterogenous Cloud Aided Multi-UAV Systems. *IEEE Trans Commun*. 2019;67(4):3026–3039.

[3] Rajashekar R, Renzo MD, Hari KVS, *et al*. A Beamforming-Aided Full-Diversity Scheme for Low-Altitude Air-to-Ground Communication Systems Operating With Limited Feedback. *IEEE Trans Commun*. 2018;66(12):6602–6613.

[4] Nguyen LD, Nguyen KK, Kortun A, *et al*. Real-Time Deployment and Resource Allocation for Distributed UAV Systems in Disaster Relief. In: *Proc. IEEE 20th International Workshop on Signal Processing Advances in Wireless Commun. (SPAWC)*. Cannes, France; 2019. p. 1–5.

[5] Xie L, Xu J, and Zhang R. Throughput Maximization for UAV-Enabled Wireless Powered Communication Networks. *IEEE Internet Things J*. 2019;6(2):1690–1703.

[6] Wu Q, Zeng Y, and Zhang R. Joint Trajectory and Communication Design for Multi-UAV Enabled Wireless Networks. *IEEE Trans Wireless Commun*. 2018;17(3):2109–2121.

[7] Nguyen MN, Nguyen LD, Duong TQ, *et al*. Real-Time Optimal Resource Allocation for Embedded UAV Communication Systems. *IEEE Wireless Commun Lett*. 2019;8(1):225–228.

[8] Liu L, Zhang S, and Zhang R. Multi-Beam UAV Communication in Cellular Uplink: Cooperative Interference Cancellation and Sum-Rate Maximization. *IEEE Trans Wireless Commun*. 2019;18(10):4679–4691.

[9] Wang H, Wang J, Ding G, Wang L, Tsiftsis TA, and Sharma PK, Resource Allocation for Energy Harvesting-Powered D2D Communication Underlaying UAV-Assisted Networks. *IEEE Trans Green Commun and Networking.* 2018;2(1):14–24.

[10] Duong TQ, Nguyen LD, Tuan HD, et al. Learning-Aided Realtime Performance Optimisation of Cognitive UAV-Assisted Disaster Communication. In: *Proc. IEEE Global Communications Conference (GLOBECOM).* Waikoloa, HI, USA; 2019.

[11] Kai C, Li H, Xu L, et al. Energy-Efficient Device-to-Device Communications for Green Smart Cities. *IEEE Trans Ind Informat.* 2018;14(4):1542–1551.

[12] Vo NS, Duong TQ, Tuan HD, et al. Optimal Video Streaming in Dense 5G Networks With D2D Communications. *IEEE Access.* 2017;6:209–223.

[13] Huang L, Bi S, and Zhang YJA. Deep Reinforcement Learning for Online Computation Offloading in Wireless Powered Mobile-Edge Computing Networks. *IEEE Trans Mobile Comput.* 2019;1–1.

[14] Nguyen KK, Duong TQ, Vien NA, et al. Distributed Deep Deterministic Policy Gradient for Power Allocation Control in D2D-Based V2V Communications. *IEEE Access.* 2019;7:164533–164543.

[15] Yu Y, Wang T, and Liew SC. Deep-Reinforcement Learning Multiple Access for Heterogeneous Wireless Networks. *IEEE J Select Areas Commun.* 2019;37(6):1277–1290.

[16] Nguyen KK, Duong TQ, Vien NA, et al. Non-Cooperative Energy Efficient Power Allocation Game in D2D Communication: A Multi-Agent Deep Reinforcement Learning Approach. *IEEE Access.* 2019;7:100480–100490.

[17] Amazon Prime Air;. Available from: https://www.amazon.com/Amazon-Prime-Air/b?ie=UTF8&node=8037720011.

[18] Vacca A, Onishi H, and Cuccu F. Drones: Military Weapons, Surveillance or Mapping Tools for Environmental Monitoring? The Need for Legal Framework Is Required. *Transp Res Proc.* 2017;25:51–62.

[19] Thiels CA, Aho JM, Zietlow SP, et al. Use of unmanned aerial vehicles for medical product transport. *Air Med J.* 2015;34(2):104–108.

[20] Duong TQ, Nguyen LD, and Nguyen LK. Practical Optimisation of Path Planning and Completion Time of Data Collection for UAV-enabled Disaster Communications. In: *Proc. 15th Int. Wireless Commun. Mobile Computing Conf. (IWCMC).* Tangier, Morocco; 2019. p. 372–377.

[21] Nguyen LD, Kortun A, and Duong TQ. An Introduction of Real-time Embedded Optimisation Programming for UAV Systems under Disaster Communication. *EAI Endorsed Trans Ind Netw Intell Systems.* 2018; 5(17):1–8.

[22] Shakoor S, Kaleem Z, Baig MI, Chughtai O, Duong TQ, and Nguyen LD, Role of UAVs in Public Safety Communications: Energy Efficiency Perspective. *IEEE Access.* 2019;7:140665–140679.

[23] Zhan C, Zeng Y, and Zhang R. Energy-Efficient Data Collection in UAV Enabled Wireless Sensor Network. *IEEE Wireless Commun Lett.* 2018;7(3):328–331.

[24] Ye H, Li GY, and Juang BHF. Deep Reinforcement Learning Based Resource Allocation for V2V Communications. *IEEE Trans Veh Technol.* 2019;68(4):3163–3173.

[25] Schulman J, Wolski F, Dhariwal P, et al. *Proximal Policy Optimization Algorithms*; 2017.

[26] Ju H and Zhang R. Throughput Maximization in Wireless Powered Communication Networks. *IEEE Trans Wireless Commun.* 2014;13(1):418–428.

[27] Al-Hourani A, Kandeepan S, and Lardner S. Optimal LAP Altitude for Maximum Coverage. *IEEE Wireless Commun Lett.* 2014;3(6):569–572.

[28] Puterman ML. *Markov Decision Processes: Discrete Stochastic Dynamic Programming*. John Wiley & Sons, Inc.; 1994.

[29] Bertsekas DP. *Dynamic Programming and Optimal Control.* vol. 1. Athena Scientific Belmont, MA; 1995.

[30] Mnih V, Kavukcuoglu K, Silver D, et al. *Playing Atari with Deep Reinforcement Learning*; 2013. Available from: https://arxiv.org/abs/1312.5602.

[31] Sutton RS, McAllester D, Singh S, et al. Policy Gradient Methods for Reinforcement Learning with Function Approximation. In: *Adv. Neural Inf. Process. Syst.*; 2000. p. 1057–1063.

[32] Kakade S. A Natural Policy Gradient. In: *Adv. Neural Inf. Process. Syst.*; 2002. p. 1531–1538.

[33] Mnih V, Badia, AP, Mirza M, et al. Asynchronous Methods for Deep Reinforcement Learning. In: *Proc. Int. Conf. Mach. Learn. PMLR*; 2016. p. 1928–1937.

[34] Lillicrap TP, Hunt JJ, Pritzel A, et al. Continuous Control with Deep Reinforcement learning. In: *Proc. 4th International Conf. on Learning Representations (ICLR)*; 2016.

[35] Schulman J, Moritz P, Levine S, et al. High-Dimensional Continuous Control Using Generalized Advantage Estimation. In: *Proc. 4th International Conf. Learning Representations (ICLR)*; 2016.

[36] Abadi M, Barham P, Chen J, et al. Tensorflow: A System for Large-Scale Machine Learning. In: *Proc. 12th USENIX Sym. Opr. Syst. Design and Imp. (OSDI 16)*; 2016. p. 265–283.

[37] Kingma DP and Ba JL. *Adam: A Method for Stochastic Optimization*; 2014. Available from: https://arxiv.org/abs/1412.6980.

[38] Diamond S and Boyd S. CVXPY: A Python-Embedded Modeling Language for Convex Optimization. *J Mach Learn Res.* 2016;17(83):2909–2913.

*Chapter 11*
# 3D trajectory design and data collection in UAV-assisted networks

Due to the limitation of UAVs' on-board power and flight time, it is challenging to obtain an optimal resource allocation scheme for the UAV-assisted Internet of Things (IoT). In this chapter, we illustrate a new UAV-assisted IoT system relying on the shortest flight path of the UAVs to maximise the amount of data collected from IoT devices. To achieve this goal, a DRL-based technique is used to find the optimal trajectory to allow the UAV to autonomously collect all the data from user nodes at a significant total sum-rate improvement while minimising the associated used resources. Numerical results are provided to highlight how such techniques strike a balance between the throughput attained, trajectory, and the time spent.[*]

## 11.1 Introduction

Given the agility of unmanned aerial vehicles (UAVs), these are capable of supporting compelling applications and are beginning to be deployed more broadly. Recently, the UK and Chile authorities proposed to deliver medical support and other essential supplies by using UAVs to vulnerable people in response to COVID-19 [2,3]. In [4], the authors used UAVs for image collection and high-resolution topography exploration. However, given the several limitations of on-board power level and the ability to adapt to changes in the environment, UAVs may not be fully autonomous and can only operate for short flight durations unless remote laser-charging is used [5]. Moreover, due to some challenging tasks such as topographic surveying, data collection, or obstacle avoidance, the existing UAV technologies cannot operate in an optimal manner.

Wireless networks supported by UAVs constitute a promising technology for enhancing network performance [6]. The applications of UAVs in wireless networks span across diverse research fields, such as wireless sensor networks [7], caching [8], heterogeneous cellular networks [9], massive multiple-input multiple-output (MIMO) [10], disaster communications [11,12], and device-to-device communications (D2D) [13]. For example, in [14], UAVs were deployed to provide network coverage for people in remote areas and disaster zones. UAVs were also used for collecting data in a wireless sensor network [7]. Nevertheless, the benefits of UAV-aided wireless communication are critically dependent on the limited on-board power

[*]This chapter has been published partly in [1].

level. Thus, the resource allocation of UAV-aided wireless networks plays a pivotal role in approaching optimal performance. Yet, the existing contributions typically assume having a static environment [11,12,15] and often ignore the stringent flight time constraints in real-life applications [7,9,16].

Machine learning has recently been proposed for the intelligent support of UAVs and other devices in the network [10,17–25]. Reinforcement learning (RL) is capable of searching for an optimal policy by trial-and-error learning. However, it is challenging for model-free RL algorithms, such as Q-learning, to obtain an optimal strategy while considering a large state and action space. Fortunately, with the emerging neural networks, the sophisticated combination of RL and DL, namely deep reinforcement learning (DRL), is eminently suitable for solving high-dimensional problems. Hence, DRL algorithms have been widely applied in fields such as robotics [26], business management [27], and gaming [28]. Recently, DRL has also become popular in solving diverse problems in wireless networks thanks to their decision-making ability and flexible interaction with the environment [8,10,19–25,29–31]. For example, DRL was used for solving problems in the areas of resource allocation [19,20,30], navigation [10,32], and interference management [23].

### 11.1.1 Related contributions

UAV-aided wireless networks have also been used for machine-to-machine communications [33] and D2D scenarios in 5G [15,34], but the associated resource allocation problems remain challenging in real-life applications. Several techniques have been developed for solving resource allocation problems [19,20,32,35–37]. In [35], the authors have conceived multi-beam UAV communications and a cooperative interference cancellation scheme for maximising the uplink sum rate received from multiple UAVs by the base stations (BS) on the ground. The UAVs were deployed as access points to serve several ground users in [36]. Then, the authors proposed successive convex programming to maximise the minimum uplink rate gleaned from all the ground users. In [32], the authors characterised the trade-off between the ground terminal transmission power and the specific UAV trajectory both in a straight and circular trajectory.

The issues of data collection, energy minimisation, and path planning have been considered in [24,33,38–46]. In [39], the authors minimised the energy consumption of the data collection task by jointly optimising the sensor nodes' wakeup schedule and the UAV trajectory. The authors of [40] proposed an efficient algorithm for joint trajectory and power allocation optimisation in UAV-assisted networks to maximise the sum rate during a specific length of time. A pair of near-optimal approaches for optimal trajectory was proposed for a given UAV power allocation and power allocation optimisation for a given trajectory. In [33], the authors introduced a communication framework for UAV-to-UAV communication under the constraints of the UAV's flight speed, location uncertainty, and communication throughput. Then, a path planning algorithm was proposed to minimise the associated completion time task while balancing the performance by computational complexity trade-off. However, these techniques mostly operate in offline modes and may impose excessive delay on the system. It is crucial to improve the decision-making time to meet the stringent requirements of UAV-assisted wireless networks.

Again, machine learning has been recognised as a powerful tool for solving the high-dynamic trajectory and resource allocation problems in wireless networks. In [37], the authors proposed a model based on the classic k-means algorithm for grouping the users into clusters and assigned a dedicated UAV to serve each cluster. By relying on their decision-making ability, DRL algorithms have been used for lending each node some degree of autonomy [8,19–22,29,30,47]. In [29], an optimal DRL-based channel access strategy to maximise the sum rate and $\alpha$-fairness was considered. In [19,20], we deployed DRL techniques for enhancing the energy efficiency of D2D communications. In [22], the authors characterised the DQL algorithm for minimising the data packet loss of UAV-assisted power transfer and data collection systems. As a further advance, caching problems were considered in [8] to maximise the cache success hit rate and to minimise the transmission delay. The authors designed both a centralised and a decentralised system model and used an actor-critic algorithm to find the optimal policy.

DRL algorithms have also been applied for path planning in UAV-assisted wireless communications [10,23–25,31,48]. In [23], the authors proposed a DRL algorithm based on the echo state network of [49] for finding the flight path, transmission power, and associated cell in UAV-powered wireless networks. The so-called deterministic policy gradient algorithm of [50] was invoked for UAV-assisted cellular networks in [31]. The UAV's trajectory was designed to maximise the uplink sum rate attained without the knowledge of the user location and the transmit power. Moreover, in [10], the authors used the DQL algorithm for the UAV's navigation based on the received signal strengths estimated by a massive MIMO scheme. In [24], Q-learning was used for controlling the movement of multiple UAVs in a pair of scenarios, namely for static user locations and for dynamic user locations under a random walk model. However, the aforementioned contributions have not addressed the joint trajectory and data collection optimisation of UAV-assisted networks, which is a difficult research challenge. Furthermore, these existing works mostly neglected interference, 3D trajectory, and dynamic environment.

## 11.1.2 Contributions and organisation

A novel DRL-aided UAV-assisted system is conceived to find the optimal UAV path for maximising the joint reward function based on the shortest flight distance and the uplink transmission rate. We boldly and explicitly contrast our proposed solution to the state-of-the-art in Tables 11.1 and 11.2. Our main contributions are further summarised as follows:

- In our UAV-aided system, the maximum amount of data is collected from the users with the shortest distance travelled.
- Our UAV-aided system is specifically designed to tackle the stringent constraints owing to the position of the destination, the UAV's limited flight time and the communication link's realistic constraints. The UAV's objective is to find the optimal trajectory for maximising the total network throughput while minimising its distance travelled.
- Explicitly, these challenges are tackled by conceiving bespoke DRL techniques for solving the above problem. To elaborate, the area is divided into a grid to

Table 11.1  A comparison with existing literature (Part 1)

|  | [38] | [7] | [22] | [24] | [41] | [10] |
|---|---|---|---|---|---|---|
| 3D trajectory |  |  |  | ✓ |  |  |
| Sum-rate maximisation | ✓ |  | ✓ | ✓ |  |  |
| Time minimisation |  | ✓ | ✓ |  |  |  |
| Dynamic environment |  |  |  | ✓ | ✓ |  |
| Unknown users |  |  |  |  |  |  |
| Reinforcement learning |  |  | ✓ | ✓ |  | ✓ |
| Deep neural networks |  |  | ✓ |  |  | ✓ |

Table 11.2  A comparison with existing literature (Part 2)

|  | [42] | [48] | [43] | [44] | Our work |
|---|---|---|---|---|---|
| 3D trajectory | ✓ |  |  | ✓ | ✓ |
| Sum-rate maximisation |  | ✓ | ✓ | ✓ | ✓ |
| Time minimisation | ✓ |  | ✓ |  | ✓ |
| Dynamic environment |  | ✓ |  |  | ✓ |
| Unknown users |  |  |  |  | ✓ |
| Reinforcement learning |  | ✓ |  |  | ✓ |
| Deep neural networks |  | ✓ |  |  | ✓ |

enable fast convergence. Following its training, the UAV can have the autonomy to make a decision concerning its next action at each position in the area, hence eliminating the need for human navigation. This makes our UAV-aided system more reliable and practical and optimises the resource requirements.
- A pair of scenarios are considered relying either on three or five clusters for quantifying the efficiency of our novel DRL techniques in terms of both the sum rate, the trajectory, and the associated time. A convincing 3D trajectory visualisation is also provided.
- Finally, but most importantly, it is demonstrated that our DRL techniques approach the performance of the optimal 'genie-solution' associated with the perfect knowledge of the environment.

Although the existing DRL algorithms have been well exploited in wireless networks, it is challenging to apply to current scenarios due to stringent constraints of the considered system, such as UAV's flying time, transmission distance, and mobile users. As with the DQL and dueling DQL algorithm, we discretise the flying path into a grid, and the UAV only needs to decide the action in a finite action space. With the finite state and action space, the neural networks can be easily trained and deployed for online phase. With other existing RL algorithm, we have tried and found out that some of them are not effective in solving our proposed problem. Meanwhile, the continuous solver RL algorithms, e.g. deep deterministic policy gradient (DDPG) and proximal policy optimisation (PPO), are not suitable and so challenging for the

# 3D trajectory design and data collection in UAV-assisted networks 165

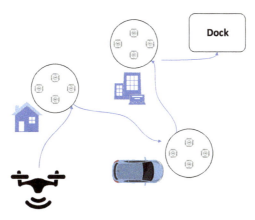

*Figure 11.1 System model of UAV-aided IoT communications. ©IEEE 2022. Reprinted with permission from [1]*

trade-off problem. Therefore, in this chapter, we propose the DQL and dueling DQL algorithm to obtain the optimal trade-off in the total achievable sum rate and trajectory. As such, we can transfer a real-life application into a digital environment for optimisation and solve it efficiently.

The rest of our chapter is organised as follows. In Section 11.2, we describe our data collection system model and the problem formulation of IoT networks relying on UAVs. Then, the mathematical background of the DRL algorithms is presented in Section 11.3. Deep Q-learning (DQL) is employed for finding the best trajectory and for solving our data collection problem in Section 11.4. Furthermore, we use the dueling DQL algorithm of [51] for improving the system performance and convergence speed in Section 11.5. Next, we characterise the efficiency of the DRL techniques in Section 11.6. Finally, in Section 11.7, we summarise our findings and discuss our future research.

## 11.2 System model and problem formulation

Consider a system consisting of a single UAV and $M$ groups of users, as shown in Figure 11.1, where the UAV relying on a single antenna visits all clusters to cover all the users. The 3D coordinate of the UAV at time step $t$ is defined as $X^t = (x_0^t, y_0^t, H_0^t)$. Each cluster consists of $K$ users, which are unknown and distributed randomly within the coverage radius of $C$. The users are moving following the random walk model with the maximum velocity $v$. The position of the $k$th user in the $m$th cluster at time step $t$ is defined as $X_{m,k}^t = (x_{m,k}^t, y_{m,k}^t)$. The UAV's objective is to find the best trajectory while covering all the users and to reach the dock upon completing its mission.

### 11.2.1 Observation model

The distance from the UAV to user $k$ in cluster $m$ at time step $t$ is given by:

$$d_{m,k}^t = \sqrt{(x_0^t - x_{m,k}^t)^2 + (y_0^t - y_{m,k}^t)^2 + H_0^{t\,2}}. \tag{11.1}$$

We assume that the communication channels between the UAV and users are dominated by line-of-sight (LoS) links; thus, the channel between the UAV and the $k$th user in the $m$th cluster at time step $t$ follows the free-space path-loss model, which is represented as:

$$h_{m,k}^t = \beta_0 d_{m,k}^{t\,-2} = \frac{\beta_0}{(x_0^t - x_{m,k}^t)^2 + (y_0 - y_{m,k}^t)^2 + H_0^{t\,2}}, \quad (11.2)$$

where the channel's power gain at a reference distance of $d = 1$ m is denoted by $\beta_0$.

The achievable throughput from the $k$th user in the $m$th cluster to the UAV at time $t$ if the user satisfies the distance constraint is defined as follows:

$$R_{m,k}^t = B \log_2 \left(1 + \frac{p_{m,k}^t h_{m,k}^t}{\sum_{i \neq m}^{M} \sum_{j}^{K} p_{i,j}^t h_{i,j}^t + \sum_{u \neq k}^{K} p_{m,u}^t h_{m,u}^t + \alpha^2}\right), \forall m, k, \quad (11.3)$$

where $B$ and $\alpha^2$ are the bandwidth and the noise power, respectively; $p_{m,k}$ is the transmit power at the $k$th user in the $m$th cluster. The total sum rate over the $T$ time step from the $k$th user in cluster $m$ to the UAV is then given by:

$$R_{m,k} = \int_0^T R_{m,k}^t dt, \forall m, k. \quad (11.4)$$

### 11.2.2 Game formulation

Both the current location and the action taken jointly influence the rewards obtained by the UAV; thus, the trial-and-error-based learning task of the UAV satisfies the Markov property. We formulate the associated Markov decision process (MDP) [52] as a four tuple $<\mathscr{S}, \mathscr{A}, \mathscr{P}_{ss'}, \mathscr{R}>$, where $\mathscr{S}$ is the state space of the UAV, $\mathscr{A}$ is the action space, $\mathscr{R}$ is the expected reward of the UAV, and $\mathscr{P}_{ss'}$ is the probability of transition from state $s$ to state $s'$, where we have $s' = s^{t+1}|s = s^t$. Through learning, the UAV can find the optimal policy $\pi^* : \mathscr{S} \to \mathscr{A}$ for maximising the reward $\mathscr{R}$. As the definition of RL, the UAV does not have any knowledge about the environment. We transfer a real-life application of the data collection in the UAV-assisted IoT networks into a digital form. Thus, the UAV only has local information, and the state is defined by the position of the UAV. We have also discretised the state and action space for learning. More particularly, we formulate the trajectory and data collection game of UAV-aided IoT networks as follows:

- *Agent*: The UAV acts like an agent interacting with the environment to find the peak of the reward.
- *State space*: We define the state space by the position of UAV as:

$$\mathscr{S} = \{x, y, H\}. \quad (11.5)$$

At time step $t$, the state of the UAV is defined as $s^t = (x^t, y^t, H^t)$.

- *Action space*: The UAV at state $s^t$ can choose an action $a^t$ of the action space by following the policy at time step $t$. By dividing the area into a grid, we can define the action space as follows:

$$\mathscr{A} = \{\text{left, right, forward, backward, upward, downward, hover}\}. \quad (11.6)$$

 The UAV moves into the environment and begins collecting information when the users are covered by the UAV. When the UAV has sufficient information $R_{m,k} \geq r_{\min}$ from the $k$th user in the $m$th cluster, that user will be marked as collected in this mission and may not be visited by the UAV again.
- *Reward function*: In joint trajectory and data collection optimisation, we design the reward function to be dependent on both the total sum rate of ground users associated with the UAV plus the reward gleaned when the UAV completes one route, which is formulated as follows:

$$R = \frac{\beta}{MK} \left( \sum_m^M \sum_k^K P(m,k) R_{m,k} \right) + \zeta R_{\text{plus}}, \quad (11.7)$$

 where $\beta$ and $\zeta$ are positive variables that represent the trade-off between the network's sum rate and UAV's movement, which will be described in the sequel. Here, $P(m,k) \in \{0,1\}$ indicates whether or not user $k$ of cluster $m$ is associated with the UAV; $R_{\text{plus}}$ is the acquired reward when the UAV completes a mission by reaching the final destination. On the other hand, the term $\frac{\sum_m^M \sum_k^K P(m,k) R_{m,k}}{MK}$ defines the average throughput of all users.
- *Probability*: We define $\mathscr{P}_{s^t s^{t+1}}(a^t, \pi)$ as the probability of transition from state $s^t$ to state $s^{t+1}$ by taking the action $a^t$ under the policy $\pi$.

At each time step $t$, the UAV chooses the action $a^t$ based on its local information to obtain the reward $r^t$ under the policy $\pi$. Then, the UAV moves to the next state $s^{t+1}$ by taking the action $a^t$ and starts collecting information from the users if any available node in the network satisfies the distance constraint. Meanwhile, the users in clusters also move to new positions following the random walk model with velocity $v$. Again, we use the DRL techniques to find the optimal policy $\pi^*$ for the UAV to maximise the reward attained (11.7). Following the policy $\pi$, the UAV forms a chain of actions $(a^0, a^1, \ldots, a^t, \ldots, a^{\text{final}})$ to reach the landing dock.

Our target is to maximise the reward expected by the UAV upon completing a single mission during which the UAV flies from the initial position over the clusters and lands at the destination. Thus, we design the trajectory reward $R_{\text{plus}}$ when the UAV reaches the destination in two different ways. First, the binary reward function is defined as follows:

$$R_{\text{plus}} = \begin{cases} 1, & X_{\text{final}} \in X_{\text{target}} \\ 0, & \text{otherwise} \end{cases}, \quad (11.8)$$

where $X_{\text{final}}$ and $X_{\text{target}}$ are the final position of the UAV and the destination, respectively. However, the UAV has to move a long distance to reach the final destination. It may also be trapped in a zone and cannot complete the mission. These situations lead to increased energy consumption and reduced convergence. Thus, we consider the value of $R_{\text{plus}}^t$ in a different form by calculating the horizontal distance between the UAV and the final destination at time step $t$, yielding:

$$R_{\text{plus}}^t = \begin{cases} 1, & X_{\text{final}} \in X_{\text{target}} \\ (\exp(d_{targ}))^{-1}, & \text{otherwise,} \end{cases} \tag{11.9}$$

where $d_{targ} = \sqrt{(x_{\text{target}} - x_0^t)^2 + (y_{\text{target}} - y_0^t)^2}$ is the distance from the UAV to the landing dock.

When we design the reward function as in (11.9), the UAV is motivated to move ahead to reach the final destination. However, one of the disadvantages is that the UAV only moves forward. Thus, the UAV is unable to attain the best performance in terms of its total sum rate in some environmental settings. We compare the performance of the two trajectory reward function definitions in Section 11.6 to evaluate the pros and cons of each approach.

In our work, we optimise the 3D trajectory of the UAV and data collection for the IoT network. Particularly, we have designed the reward function as a trade-off game between the achievable sum rate and the trajectory. Denote the flying path of the UAV from the initial point to the final position by $X = (X_0, X_1, \ldots, X_{\text{final}})$; the agent needs to learn by interacting with the environment to find an optimal $X$. We have defined a trade-off value $\beta$ and $\zeta$ to make our approach more adaptive and flexible. By modifying the value of $\beta/\zeta$, the UAV adapts to several scenarios: (a) fast deployment for emergency services, (b) maximising the total sum rate, and (c) maximising the number of connections between the UAV and users. Depending on the specific problems, we can adjust the value of the trade-off parameters $\beta, \zeta$ to achieve the best performance. Thus, the game formulation is defined as follows:

$$\max R = \frac{\beta}{MK} \left( \sum_m^M \sum_k^K P(m,k) R_{m,k} \right) + \zeta R_{\text{plus}},$$

$$\text{s.t.} \quad X_{\text{final}} = X_{\text{target}},$$
$$d_{m,k} \leq d_{\text{cons}},$$
$$R_{m,k} \geq r_{\text{min}}, \tag{11.10}$$
$$P(m,k) \in \{0,1\},$$
$$T \leq T_{\text{cons}}$$
$$\beta \geq 0, \zeta \geq 0,$$

where $T$ and $T_{\text{cons}}$ are the number of steps that the UAV takes in a single mission and the maximum number of UAV's steps given its limited power, respectively. The term $X_{\text{final}} = X_{\text{target}}$ denotes the completed flying route when the final position of the UAV belongs to the destination zone. We have designed the reward function following this constraint with two functions: binary reward function in (11.8) and exponential reward function in (11.9). The term $d_{m,k} \leq d_{\text{cons}}, R_{m,k} \geq r_{\text{min}}, P(m,k) \in \{0,1\}$ denote the communication constraint. Particularly, the distance constraint $d_{m,k} \leq d_{\text{cons}}$ indicates that the served $(m,k)$-user has a satisfying distance to the UAV. $P(m,k) \in \{0,1\}$ indicates whether or not user $k$ of cluster $m$ is associated with the UAV. $R_{m,k} \geq r_{\text{min}}$ denotes the minimum information collected during the flying path. All the constraints are integrated into the reward functions in the RL algorithm. The term $T \leq T_{\text{cons}}$ denotes the constraint about the flying time. Considering the maximum flying time is $T_{\text{cons}}$, the UAV needs to complete a route by reaching the destination zone before $T_{\text{cons}}$. If the UAV can not complete a route before $T_{\text{cons}}$, the $R_{\text{plus}} = 0$ as we defined in (11.8) and (11.9). We have the trade-off value in reward function $\beta \geq 0$, $\zeta \geq 0$. Those stringent constraints, such as the transmission distance, position, and flight time, make the optimisation problem more challenging. Thus, we propose DRL techniques for the UAV in order to attain optimal performance.

## 11.3 Preliminaries

In this section, we introduce the fundamental concept of Q-learning, where the so-called value function is defined by a reward of the UAV at state $s^t$ as:

$$V(s, \pi) = \mathbb{E}\left[\sum_{t}^{T} \gamma \mathscr{R}^t(s^t, \pi) | s_0 = s\right], \quad (11.11)$$

where $\mathbb{E}[\,\cdot\,]$ represents an average of the number of samples, and $0 \leq \gamma \leq 1$ denotes the discount factor. In a finite game, there is always an optimal policy $\pi^*$ that satisfies the Bellman optimality equation [53]:

$$V^*(s, \pi) = V(s, \pi^*)$$
$$= \max_{a \in \mathscr{A}} \left[\mathbb{E}\left[\mathscr{R}^t(s^t, \pi^*)\right] + \gamma \sum_{s' \in S} \mathscr{P}_{ss'}(a, \pi^*) V(s', \pi^*)\right]. \quad (11.12)$$

The action-value function is obtained when the agent at state $s^t$ takes action $a^t$ and receives the reward $r^t$ under the agent policy $\pi$. The optimal Q-value can be formulated as follows:

$$Q^*(s, a, \pi) = \mathbb{E}\left[\mathscr{R}^t(s^t, \pi^*)\right] + \gamma \sum_{s' \in S} \mathscr{P}_{ss'}(a, \pi^*) V(s', \pi^*). \quad (11.13)$$

The optimal policy $\pi^*$ can be obtained from $Q^*(s, a, \pi)$ as follows:

$$V^*(s, \pi) = \max_{a \in \mathscr{A}} Q(s, a, \pi). \quad (11.14)$$

From (11.13) and (11.14), we have:

$$Q^*(s, a, \pi) = \mathbb{E}\left[\mathscr{R}^t(s^t, \pi^*)\right] + \gamma \sum_{s' \in S} \mathscr{P}_{ss'}(a, \pi^*) \max_{a' \in \mathscr{A}} Q(s', a', \pi),$$
$$= \mathbb{E}\left[\mathscr{R}^t(s^t, \pi^*) + \gamma \max_{a' \in \mathscr{A}} Q(s', a', \pi)\right], \quad (11.15)$$

where the agent takes the action $a' = a^{t+1}$ at state $s^{t+1}$.

Through learning, the Q-value is updated based on the available information as follows:

$$Q(s, a, \pi) = Q(s, a, \pi)$$
$$+ \alpha \left[\mathscr{R}^t(s^t, \pi^*) + \gamma \max_{a' \in \mathscr{A}} Q(s', a', \pi) - Q(s, a, \pi)\right], \quad (11.16)$$

where $\alpha$ denotes the updated parameter of the Q-value function.

In RL algorithms, it is challenging to balance the *exploration* and *exploitation* for appropriately selecting the action. The most common approach relies on the $\varepsilon$-greedy policy for the action selection mechanism as follows:

$$a = \begin{cases} \text{argmax}\, Q(s, a, \pi) & \text{with } \varepsilon \\ \text{randomly} & \text{if } 1 - \varepsilon. \end{cases} \quad (11.17)$$

Upon assuming that each episode lasts $T$ steps, the action at time step $t$ is $a^t$ that is selected by following the $\varepsilon$-greedy policy as in (11.17). The UAV at state $s^t$ communicates with the user nodes from the ground if the distance constraint of $d_{m,k} \leq d_{\text{cons}}$ is satisfied. Following the information transmission phase, the user nodes are marked as collected users and may not be revisited later during that mission. Then, after obtaining the immediate reward $r(s^t, a^t)$, the agent at state $s^t$ takes action $a^t$ to move to state $s^{t+1}$ as well as to update the $Q$-value function in (11.16). Each episode ends when the UAV reaches the final destination, and the flight-duration constraint is satisfied.

## 11.4 An effective deep reinforcement learning approach for UAV-assisted IoT networks

In this section, we conceive the DQL algorithm for trajectory and data collection optimisation in UAV-aided IoT networks. However, the Q-learning technique typically falters for large state and action spaces due to its excessive Q-table size. Thus, instead of applying the Q-table in Q-learning, we use deep neural networks to represent the relationship between the action and state space. Furthermore, we employ a pair of techniques for stabilising the neural network's performance in our DQL algorithm as follows:

- *Experience replay buffer*: Instead of using current experience, we use a so-called replay buffer $\mathscr{B}$ to store the transitions $(s, a, r, s')$ for supporting the neural network in overcoming any potential instability. When the buffer $\mathscr{B}$ is filled with the transitions, we randomly select a mini-batch of $K$ samples for training the networks. The finite buffer size of $\mathscr{B}$ allows it to be always up-to-date, and the neural networks learn from the new samples.
- *Target networks*: If we use the same network to calculate the state-action value $Q$ and the target network, the network can be shifted dramatically in the training phase. Thus, we employ a target network $Q'$ for the target value estimator. After a number of iterations, the parameters of the target network $Q'$ will be updated by the network $Q$.

The UAVs start from the initial position and interact with the environment to find the proper action in each state. The agent chooses the action $a^t$ following the current policy at state $s^t$. By executing the action $a^t$, the agent receives the response from the environment with reward $r^t$ and new state $s^{t+1}$. After each time step, the UAVs have new positions, and the environment is changed with moving users. The obtained transitions are stored in a finite memory buffer and used for training the neural networks.

The neural network parameters are updated by minimising the loss function defined as follows:

$$\mathbb{L}(\theta) = \mathbb{E}_{s,a,r,s'}\left[\left(y^{\mathrm{DQL}} - Q(s,a;\theta)\right)^2\right], \tag{11.18}$$

where $\theta$ is a parameter of the network $Q$ and we have:

$$y = \begin{cases} r^t & \text{if terminated at } s^{t+1} \\ r^t + \gamma \max_{a' \in \mathscr{A}} Q'(s', a'; \theta') & \text{otherwise.} \end{cases} \tag{11.19}$$

The details of the DQL approach in our joint trajectory and data collection trade-off game designed for UAV-aided IoT networks are presented in Algorithm 11.1 where $L$ denotes the number of episodes. Moreover, in this chapter, we design the reward obtained in each step to assume one of two different forms and compare them in our simulation results. First, the difference between the current and previous

---

**Algorithm 11.1** The deep Q-learning algorithm for trajectory and data collection optimisation in UAV-aided IoT networks

---
1: Initialise the network $Q$ and the target network $Q'$ with the random parameters $\theta$ and $\theta'$, respectively
2: Initialise the replay memory pool $\mathscr{B}$
3: **for** episode = 1, ..., $L$ **do**
4:     Receive initial observation state $s^0$
5:     **while** $X_{\mathrm{final}} \notin X_{\mathrm{target}}$ or $T \leq T_{\mathrm{cons}}$ **do**
6:         Obtain the action $a^t$ of the UAV according to the $\varepsilon$-greedy mechanism (11.17)
7:         Execute the action $a^t$ and estimate the reward $r^t$ according to (11.7)
8:         Observe the next state $s^{t+1}$
9:         Store the transition $(s^t, a^t, r^t, s^{t+1})$ in the replay buffer $\mathscr{B}$
10:        Randomly select a mini-batch of $K$ transitions $(s^k, a^k, r^k, s^{k+1})$ from $\mathscr{B}$
11:        Update the network parameters using gradient descent to minimise the loss

$$\mathbb{L}(\theta) = \mathbb{E}_{s,a,r,s'}\left[\left(y^{\mathrm{DQL}} - Q(s,a;\theta)\right)^2\right], \tag{11.20}$$

The gradient update is

$$\nabla_\theta \mathbb{L}(\theta) = \mathbb{E}_{s,a,r,s'}\left[\left(y^{\mathrm{DQL}} - Q(s,a;\theta)\right)\nabla_\theta Q(s,a;\theta)\right], \tag{11.21}$$

12:        Update the state $s^t = s^{t+1}$
13:        Update the target network parameters after a number of iterations as $\theta' = \theta$
14:     **end while**
15: **end for**

reward of the UAV is:

$$r_1^t(s^t, a^t) = r^t(s^t, a^t) - r^{t-1}(s^{t-1}, a^{t-1}). \tag{11.22}$$

Second, we design the total episode reward as the accumulation of all immediate rewards of each step within one episode as:

$$r_2^t(s^t, a^t) = \sum_{i=0}^{t} r_1^t(s^t, a^t). \tag{11.23}$$

## 11.5 Deep reinforcement learning approach for UAV-assisted IoT networks: A dueling deep Q-learning approach

In [51], the standard Q-learning algorithm often falters due to the oversupervision of all the state-action pairs. In addition, it is unnecessary to estimate the value of each action choice in a particular state. For example, in our environment setting, the UAV has to consider moving either to the left or to the right when hitting the boundaries. Thus, we can improve the convergence speed by avoiding visiting all state-action pairs. Instead of using the Q-value function of the conventional DQL algorithm, the dueling neural network of [51] is introduced to improve the convergence rate and stability. The advantage function $A(s, a) = Q(s, a) - V(s)$ related both to the value function and the Q-value function describes the importance of each action related to each state.

The idea of a dueling deep network is based on a combination of two streams of the value function and the advantage function used for estimating the single output Q-function. One of the streams of a fully connected layer estimates the value function $V(s; \theta_V)$, while the other stream outputs a vector $A(s, a; \theta_A)$, where $\theta_A$ and $\theta_V$ represent the parameters of the two networks. The Q-function can be obtained by combining the two streams' outputs as

$$Q(s, a; \theta, \theta_A, \theta_V) = V(s; \theta_V) + A(s, a; \theta_A). \tag{11.24}$$

Equation (11.24) applies to all $(s, a)$ instances; thus, we have to replicate the scalar $V(s; \theta_V)$, $|\mathcal{A}|$ times to form a matrix. However, $Q(s, a; \theta, \theta_A, \theta_V)$ is a parameterised estimator of the true Q-function; thus, we cannot uniquely recover the value function $V$ and the advantage function $A$. Therefore, (11.27) results in poor practical performances when used directly. To address this problem, we can map the advantage function estimator to have no advantage at the chosen action by combining the two streams as:

$$Q(s, a; \theta, \theta_A, \theta_V) = V(s; \theta_V) + \left( A(s, a; \theta_A) - \max_{a' \in |\mathcal{A}|} A(s, a'; \theta_A) \right). \tag{11.25}$$

For $a^* = \mathrm{argmax}_{a' \in \mathcal{A}} Q(s, a'; \theta, \theta_A, \theta_V) = \mathrm{argmax}_{a' \in \mathcal{A}} A(s, a'; \theta_A)$, we have $Q(s, a^*; \theta, \theta_A, \theta_V) = V(s; \theta_V)$. Hence, the stream $V(s; \theta_V)$ estimates the value function while the other stream is the advantage function estimator. We can transform (11.25) using an average formulation instead of the *max* operator as follows:

$$Q(s, a; \theta, \theta_A, \theta_V) = V(s; \theta_V) + \left( A(s, a; \theta_A) - \frac{1}{|\mathscr{A}|} \sum_{a'} A(s, a'; \theta_A) \right). \quad (11.26)$$

Now, we can solve the problem of identifiability by subtracting the mean as in (11.26). Based on (11.26), we propose a dueling DQL algorithm for our joint trajectory and data collection problem in UAV-assisted IoT networks relying on Algorithm 11.2. Note that estimating $V(s; \theta_V)$ and $A(s, a; \theta_A)$ does not require any extra supervision, and they will be computed automatically.

---

**Algorithm 11.2** The dueling deep Q-learning algorithm for trajectory and data collection optimisation in UAV-aided IoT networks

---

1: Initialise the network $Q$ and the target network $Q'$ with the random parameters, $\theta$ and $\theta'$, respectively
2: Initialise the replay memory pool $\mathscr{B}$
3: **for** episode = $1, \ldots, L$ **do**
4:     Receive the initial observation state $s^0$
5:     **while** $X_{\text{final}} \notin X_{\text{target}}$ or $T \leq T_{\text{cons}}$ **do**
6:         Obtain the action $a^t$ of the UAV according to the $\varepsilon$-greedy mechanism (11.17)
7:         Execute the action $a^t$ and estimate the reward $r^t$ according to (11.7)
8:         Observe the next state $s^{t+1}$
9:         Store the transition $(s^t, a^t, r^t, s^{t+1})$ in the replay buffer $\mathscr{B}$
10:        Randomly select a mini-batch of $K$ transitions $(s^k, a^k, r^k, s^{k+1})$ from $\mathscr{B}$
11:        Estimate the $Q$-value function by combining the two streams as follows:

$$Q(s, a; \theta, \theta_A, \theta_V) = V(s; \theta_V)$$
$$+ \left( A(s, a; \theta_A) - \frac{1}{|\mathscr{A}|} \sum_{a'} A(s, a'; \theta_A) \right). \quad (11.27)$$

12:        Update the network parameters using gradient descent to minimise the loss

$$\mathbb{L}(\theta) = \mathbb{E}_{s,a,r,s'} \left[ \left( y^{\text{DuelingDQL}} - Q(s, a; \theta, \theta_A, \theta_V) \right)^2 \right], \quad (11.28)$$

13:        where

$$y^{\text{DuelingDQL}} = r^t + \gamma \max_{a' \in \mathscr{A}} Q'(s', a'; \theta', \theta_A, \theta_V). \quad (11.29)$$

14:        Update the state $s^t = s^{t+1}$
15:        Update the target network parameters after a number of iterations as $\theta' = \theta$
16:     **end while**
17: **end for**

Table 11.3 Simulation parameters

| Parameters | Value |
| --- | --- |
| Bandwidth ($W$) | 1 MHz |
| UAV transmission power | 5 W |
| The start position of UAV | (0, 0, 200) |
| Discounting factor | $\gamma = 0.9$ |
| Max number of users per cluster | 10 |
| Noise power | $\alpha^2 = -110$ dBm |
| The reference channel power gain | $\beta_0 = -50$ dB |
| Path-loss exponent | 2 |

## 11.6 Simulation results

In this section, we present our simulation results characterising the joint optimisation problem of UAV-assisted IoT networks. To highlight the efficiency of our proposed model and the DRL methods, we consider a pair of scenarios: a simple one having three clusters and a more complex one with five clusters in the coverage area. We use Tensorflow 1.13.1 [54], and the Adam optimiser of [55] for training the neural networks. We set the maximum value of $\beta/\zeta$ not too large because we prefer the completion of a mission. The maximum value is set to $\beta/\zeta = 4/1$. The other parameters are provided in Table 11.3.

In Figure 11.2, we present the trajectory obtained after training using the DQL algorithm in the 5-cluster scenario. The green circle and blue dots represent the clusters' coverage and the user nodes, respectively. The red line and black line in the figure represent the UAV's trajectory based on our method in (11.8) and (11.9), respectively. The UAV starts at (0, 0), visits about 40 users, and lands at the destination that is denoted by the black star. In a complex environment setting, it is challenging to expect the UAV to visit all users while satisfying the flight-duration and power level constraints.

### 11.6.1 Expected reward

We compare our proposed algorithm with optimal performance and the Q-learning algorithm. However, to achieve the optimal results, we have defined some assumptions about the IoT's position and the unlimited power level of the UAV. For purposes of comparison, we run the algorithm five times in five different environmental settings and take the average to draw the figures. First, we compare the reward obtained following (11.7). Let us consider the 3-cluster scenario and $\beta/\zeta = 2 : 1$ in Figure 11.3a, where the DQL and the dueling DQL algorithms using the exponential function (11.9) reach the best performance. When using the exponential trajectory design function (11.9), the performance converges faster than that of the DQL and of the dueling DQL methods using the binary trajectory function (11.8). The performance

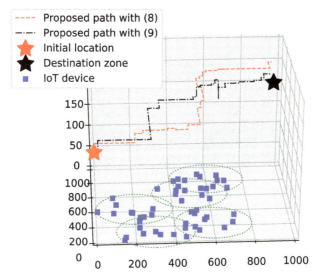

*Figure 11.2  Trajectory obtained by using our dueling DQL algorithm. ©IEEE 2022. Reprinted with permission from [1]*

of using the Q-learning algorithm is the worst. In addition, in Figure 11.3b, we compare the performance of the DQL and dueling DQL techniques using different $\beta/\zeta$ values. The average performance of the dueling DQL algorithm is better than that of the DQL algorithm. Furthermore, the results of using the exponential function (11.9) are better than those of the ones using the binary function (11.8). When the value of $\beta/\zeta \geq 1:2$, the performance achieved by the DQL and dueling DQL algorithm is close to the optimal performance.

Furthermore, we compare the rewards obtained by the DQL and dueling DQL algorithms in complex scenarios with 5 clusters and 50 user nodes in Figure 11.4. The performance of using the episode reward (11.23) is better than that of using the immediate reward (11.22) in both trajectory designs relying on the DQL and dueling DQL algorithms. In Figure 11.4a, we compare the performance in conjunction with the binary trajectory design, while in Figure 11.4b, the exponential trajectory design is considered. For $\beta/\zeta = 1:1$, the rewards obtained by the DQL and dueling DQL are similar and stable after about 400 episodes. When using the exponential function (11.9), the dueling DQL algorithm reaches the best performance and is close to the optimal performance. Moreover, the convergence of the dueling DQL technique is faster than that of the DQL algorithm. In both reward definitions, the Q-learning with (11.22) shows the worst performance.

In Figure 11.5, we compare the performance of the DQL and the dueling DQL algorithms while considering different $\beta/\zeta$ parameter values. The dueling DQL algorithm shows better performance for all the $\beta/\zeta$ pair values, exhibiting better rewards. Additionally, when using the exponential function (11.9), both proposed algorithms show better performance than the ones using the binary function (11.8) if $\beta/\zeta \leq 1:1$, but it becomes less effective when $\beta/\zeta$ is set higher. Again, we achieve

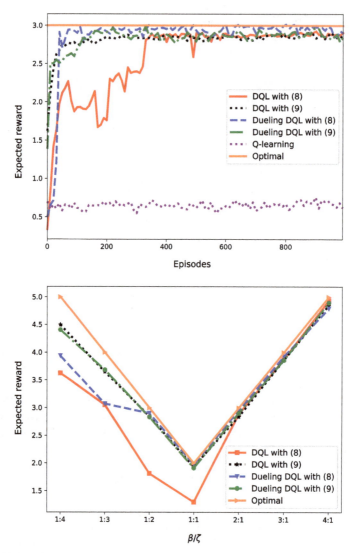

*Figure 11.3   The performance when using the DQL and dueling DQL algorithms with three clusters while considering different $\beta/\zeta$ values. ©IEEE 2022. Reprinted with permission from [1]*

a near-optimal solution while we consider a complex environment without knowing the IoT nodes' position and mobile users. It is challenging to expect the UAV to visit all IoT nodes with limited flying power and duration.

We compare the performance of the DQL and the dueling DQL algorithm using different reward function settings in Figures 11.6 and 11.7, respectively. The DQL algorithm reaches the best performance when using the episode reward (11.23) in

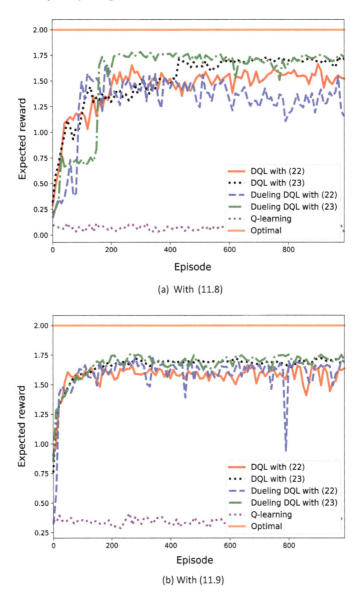

Figure 11.4 The expected reward when using the DQL and dueling DQL algorithms with 5-cluster scenario. ©IEEE 2022. Reprinted with permission from [1]

Figure 11.6, while the fastest convergence speed can be achieved by using the exponential function (11.9). When $\beta/\zeta \geq 1 : 1$, the DQL algorithm relying on the episode function (11.23) outperforms the ones using the immediate reward function (11.22) in Figure 11.6b. The reward (11.7) using the exponential trajectory design (11.9) has

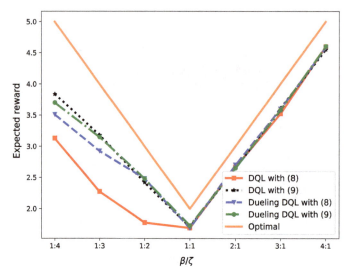

*Figure 11.5  The performance when using the DQL and dueling DQL algorithms with 5 clusters and different $\beta/\zeta$ values. ©IEEE 2022. Reprinted with permission from [1]*

a better performance than that using the binary trajectory design (11.8) for all the $\beta/\zeta$ values. Similar results are shown when using the dueling DQL algorithm in Figure 11.7. The immediate reward function (11.22) is less effective than the episode reward function (11.23).

### 11.6.2  Throughput comparison

In (11.7), we consider two elements: the trajectory cost and the average throughput. In order to quantify the communication efficiency, we compare the total throughput in different scenarios. In Figure 11.8, the performances of the DQL algorithm associated with several $\beta/\zeta$ values are considered while using the binary trajectory function (11.8), the episode reward (11.23) and 3 clusters. The throughput obtained for $\beta/\zeta = 1:1$ is higher than that of the others, and when $\beta$ increases, the performance degrades. However, when comparing with Figure 11.3b, we realise that in some scenarios, the UAV was stuck and could not find the way to the destination. That leads to increased flight time spent and distance travelled. More details are shown in Figure 11.8b, where we compare the expected throughput of both the DQL and dueling DQL algorithms. The best throughput is achieved when using the dueling DQL algorithm with $\beta/\zeta = 1:1$ in conjunction with (11.8), which is higher than the peak of the DQL method with $\beta/\zeta = 1:2$.

In Figure 11.9, we compare the throughput of different techniques in the 5-cluster scenario. Let us now consider the binary trajectory design function (11.8) in Figure 11.9a, where the DQL algorithm achieves the best performance using

*3D trajectory design and data collection in UAV-assisted networks* 179

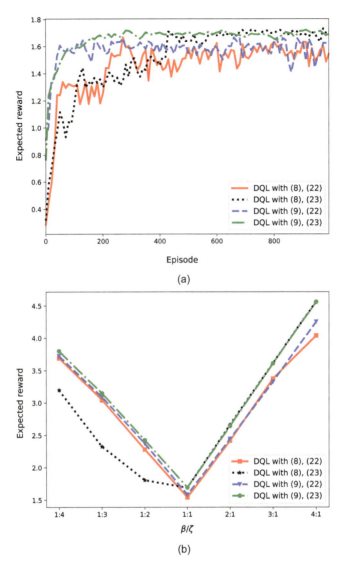

Figure 11.6 *The expected reward when using the DQL algorithm with 5 clusters and different reward function settings. ©IEEE 2022. Reprinted with permission from [1]*

$\beta/\zeta = 1 : 1$ and $\beta/\zeta = 2 : 1$. There is a slight difference between the DQL method having different settings when using exponential the trajectory design function (11.9), as shown in Figure 11.9b.

In Figures 11.10 and 11.11, we compare the throughput of different $\beta/\zeta$ pairs. The DQL algorithm reaches the optimal throughput with the aid of trial-and-learn

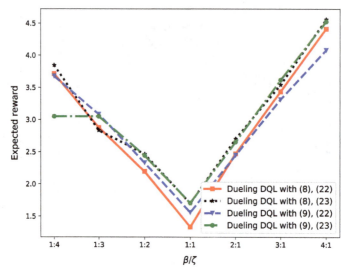

Figure 11.7 The performance when using the dueling DQL with 5 clusters and different $\beta/\zeta$ values. ©IEEE 2022. Reprinted with permission from [1]

methods. Hence, it is important to carefully design the reward function to avoid excessive offline training. As shown in Figure 11.10, the DQL and dueling DQL algorithm exhibit reasonable stability for several $\beta/\zeta \leq 1:1$ pairs and reward functions. While we achieve a similar expected reward with different reward settings in Figure 11.6, the throughput is degraded when $\beta/\zeta$ increases. In contrast, with higher $\beta$ values, the UAV can finish the mission faster. It is a trade-off game when we can choose an approximate $\beta/\zeta$ value for our specific purposes. When we employ the DQL and the dueling DQL algorithms with the episode reward (11.23), the throughput attained is higher than that using the immediate reward (11.22) with different $\beta/\zeta$ values.

Furthermore, we compare the expected throughput of the DQL and the dueling DQL algorithm when using the exponential trajectory design (11.9) in Figure 11.11a and episode reward (11.23) in Figure 11.11b. In Figure 11.11a, the dueling DQL method outperforms the DQL algorithm for almost all $\beta/\zeta$ values in both function (11.22) and (11.23). When we use the episode reward (11.23), the obtained throughput is stable with different $\beta/\zeta$ values. The throughput attained by using the exponential function (11.9) is lower than that using the binary trajectory (11.8) and the episode reward (11.23) is higher than that using the immediate reward (11.22). We can achieve the best performance when using the dueling DQL algorithm with (11.9) and (11.23). However, in some scenarios, we achieve better performance with different algorithmic settings, as we can see in Figures 11.8b and 11.10a. Thus, there is a trade-off governing the choice of the algorithm and function design.

*3D trajectory design and data collection in UAV-assisted networks* 181

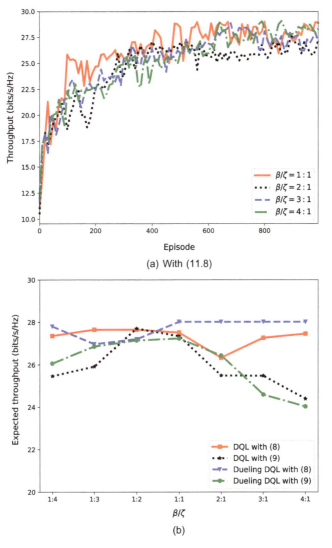

*Figure 11.8  The network's sum rate when using the DQL and dueling DQL algorithms with 3 clusters. ©IEEE 2022. Reprinted with permission from [1]*

## 11.6.3  Parametric study

In Figure 11.12, we compare the performance of our DQL technique using different *exploration* parameters $\gamma$ and $\varepsilon$ values in our $\varepsilon$-greedy method. The DQL algorithm achieves the best performance with the discounting factor of $\gamma = 0.9$ and $\varepsilon = 0.9$ in the 5-cluster scenario of Figure 11.12. Balancing the *exploration* and *exploitation* as well as the action chosen is quite challenging to maintain a steady performance of

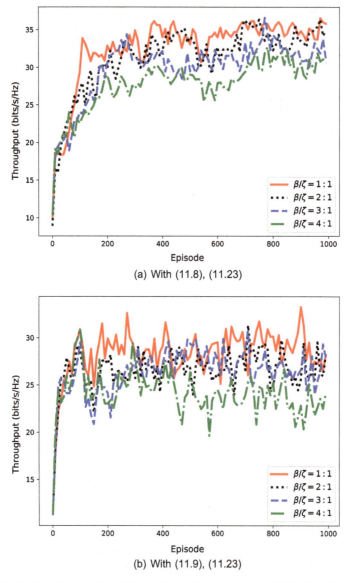

*Figure 11.9 The obtained total throughput when using the DQL algorithm with 5 clusters. ©IEEE 2022. Reprinted with permission from [1]*

the DQL algorithm. Based on the results of Figure 11.12, we opted for $\gamma = 0.9$ and $\varepsilon = 0.9$ for our algorithmic setting.

Next, we compare the expected reward of different mini-batch sizes, $K$. In the 5-cluster scenario of Figure 11.13, the DQL achieves the optimal performance with a batch size of $K = 32$. There is a slight difference in terms of convergence speed, with

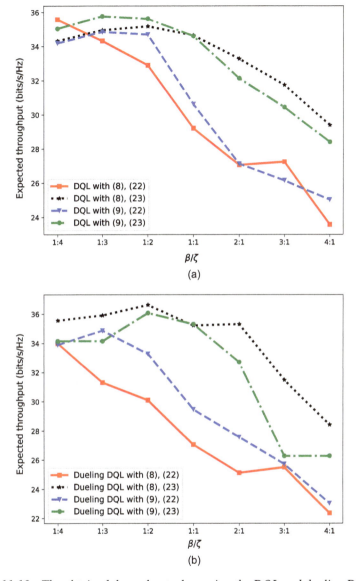

*Figure 11.10  The obtained throughput when using the DQL and dueling DQL algorithms in a 5-cluster scenario. ©IEEE 2022. Reprinted with permission from [1]*

batch size $K = 32$ being the fastest. Overall, we set the mini-batch size to $K = 32$ for our DQL algorithm.

Figure 11.14 shows the performance of the DQL algorithm with different learning rates in updating the neural network parameters while considering the scenarios of 5 clusters. When the learning rate is as high as $\alpha = 0.01$, the pace of updating

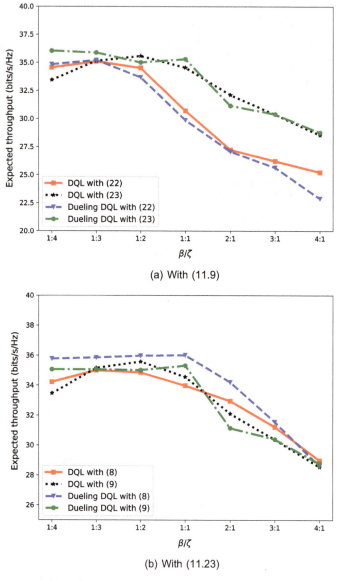

Figure 11.11 *The expected throughput when using the DQL and dueling DQL algorithms with 5 clusters. ©IEEE 2022. Reprinted with permission from [1]*

the network may result in fluctuating performance. Moreover, when $\alpha = 0.0001$ or $\alpha = 0.00001$, the convergence speed is slower and may be stuck in a local optimum instead of reaching the global optimum. Thus, based on our experiments, we opted for the learning rate of $\alpha = 0.001$ for the algorithms.

*3D trajectory design and data collection in UAV-assisted networks* 185

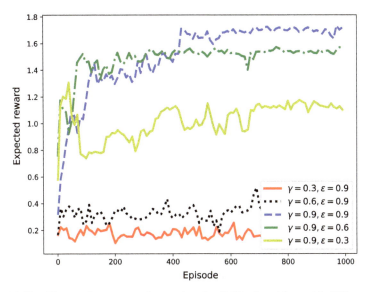

*Figure 11.12* The performance when using the DQL algorithm with different discount factors, $\gamma$, and exploration factors, $\varepsilon$. ©IEEE 2022. Reprinted with permission from [1]

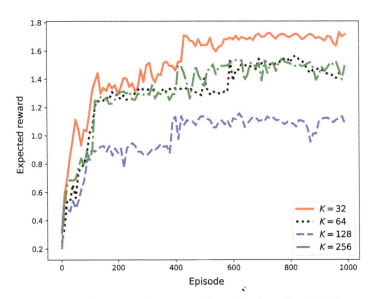

*Figure 11.13* The performance when using the DQL algorithm in 5-cluster scenario and different batch sizes, $K$. ©IEEE 2022. Reprinted with permission from [1]

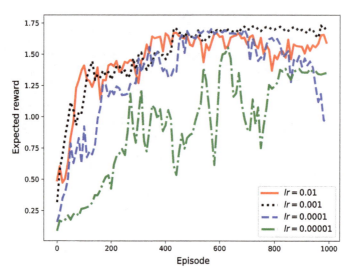

*Figure 11.14 The performance when using DQL algorithm with different learning rate, lr. ©IEEE 2022. Reprinted with permission from [1]*

## 11.7 Summary

In this chapter, the DRL technique has been proposed to jointly optimise the flight trajectory and data collection performance of UAV-assisted IoT networks. In particular, an optimisation game has been formulated to balance the flight time and total throughput while guaranteeing the quality-of-service constraints, which have been solved through a DRL-based technique. Such a DRL-based technique was able to take into account the limited UAV power level and the associated communication constraints. Simulation results showed the efficiency of such a technique, implemented through DQL and the dueling DQL approach, as well as its low computational complexity.

## References

[1] Nguyen KK, Duong TQ, Do-Duy T, *et al.* 3D UAV Trajectory and Data Collection Optimisation Via Deep Reinforcement Learning. *IEEE Trans Commun.* 2022;70(4):2358–2371.

[2] *Drone Trial to Help Isle of Wight Receive Medical Supplies Faster during COVID19 Pandemic.* Available from: https://www.southampton.ac.uk/news/2020/04/drones-covid-iow.page.

[3] *This Chilean Community Is Using Drones to Deliver Medicine to the Elderly.* Available from: https://www.weforum.org/agenda/2020/04/drone-chile-covid19/.

[4] Gao M, Xu X, Klinger Y, *et al.* High-Resolution Mapping Based on an Unmanned Aerial Vehicle (UAV) to Capture Paleoseismic Offsets Along the Altyn-Tagh Fault, China. *Sci Rep.* 2017;7(1):1–11.

[5] Liu Q, Wu J, Xia P, *et al.* Charging Unplugged: Will Distributed Laser Charging for Mobile Wireless Power Transfer Work? *IEEE Veh Technol Mag.* 2016;11(4):36–45.

[6] Claussen H. Distributed Algorithms for Robust Self-deployment and Load Balancing in Autonomous Wireless Access Networks. *In: Proc. IEEE Int. Conf. on Commun. (ICC).* vol. 4. Istanbul, Turkey; 2006. p. 1927–1932.

[7] Gong J, Chang TH, Shen C, *et al.* Flight Time Minimization of UAV for Data Collection Over Wireless Sensor Networks. *IEEE J Select Areas Commun.* 2018;36(9):1942–1954.

[8] Zhong C, Gursoy MC and Velipasalar S. Deep Reinforcement Learning-Based Edge Caching in Wireless Networks. *IEEE Trans Cogn Commun Netw.* 2020;6(1):48–61.

[9] Wu H, Wei Z, Hou Y, *et al.* Cell-Edge User Offloading via Flying UAV in Non-Uniform Heterogeneous Cellular Networks. *IEEE Trans Wireless Commun.* 2020;19(4):2411–2426.

[10] H. Huang, Y. Yang, H. Wang, Z. Ding, H. Sari, and F. Adachi, "Deep reinforcement learning for UAV navigation through massive MIMO technique," *IEEE Trans. Veh. Technol.*, vol. 69, no. 1, pp. 1117–1121, Jan. 2020.

[11] Duong TQ, Nguyen LD, Tuan HD, *et al.* Learning-Aided Realtime Performance Optimisation of Cognitive UAV-Assisted Disaster Communication. *In: Proc. IEEE Global Communications Conference (GLOBECOM).* Waikoloa, HI, USA; 2019.

[12] Duong TQ, Nguyen LD, and Nguyen LK. Practical Optimisation of Path Planning and Completion Time of Data Collection for UAV-enabled Disaster Communications. *In: Proc. 15th Int. Wireless Commun. Mobile Computing Conf. (IWCMC).* Tangier, Morocco; 2019. p. 372–377.

[13] Mozaffari M, Saad W, Bennis M, *et al.* Unmanned Aerial Vehicle With Underlaid Device-to-Device Communications: Performance and Tradeoffs. *IEEE Trans Wireless Commun.* 2016;15(6):3949–3963.

[14] Nguyen LD, Kortun A, and Duong TQ. An Introduction of Real-time Embedded Optimisation Programming for UAV Systems under Disaster Communication. *EAI Endorsed Trans Ind Netw Intell Syst.* 2018;5(17):1–8.

[15] Nguyen MN, Nguyen LD, Duong TQ, *et al.* Real-Time Optimal Resource Allocation for Embedded UAV Communication Systems. *IEEE Wireless Commun Lett.* 2019;8(1):225–228.

[16] Li X, Yao H, Wang J, *et al.* A Near-Optimal UAV-Aided Radio Coverage Strategy for Dense Urban Areas. *IEEE Trans Veh Technol.* 2019;68(9):9098–9109.

[17] Zhang H and Hanzo L. Federated Learning Assisted Multi-UAV Networks. *IEEE Trans Veh Technol.* 2020;69(11):14104–14109.

[18] Liu X, Liu Y, Chen Y, *et al.* Trajectory Design and Power Control for Multi-UAV Assisted Wireless Networks: A Machine Learning Approach. *IEEE Trans Veh Technol.* 2019;68(8):7957–7969.

[19] Nguyen KK, Duong TQ, Vien NA, et al. Distributed Deep Deterministic Policy Gradient for Power Allocation Control in D2D-Based V2V Communications. *IEEE Access*. 2019;7:164533–164543.

[20] Nguyen KK, Duong TQ, Vien NA, et al. Non-Cooperative Energy Efficient Power Allocation Game in D2D Communication: A Multi-Agent Deep Reinforcement Learning Approach. *IEEE Access*. 2019;7:100480–100490.

[21] Nguyen KK, Vien NA, Nguyen LD, et al. Real-Time Energy Harvesting Aided Scheduling in UAV-Assisted D2D Networks Relying on Deep Reinforcement Learning. *IEEE Access*. 2021;9:3638–3648.

[22] Li K, Ni W, Tovar E, et al. On-Board Deep Q-Network for UAV-Assisted Online Power Transfer and Data Collection. *IEEE Trans Veh Technol*. 2019;68(12):12215–12226.

[23] Challita U, Saad W, and Bettstetter C. Interference Management for Cellular-Connected UAVs: A Deep Reinforcement Learning Approach. *IEEE Trans Wireless Commun*. 2019;18(4):2125–2140.

[24] Liu X, Liu Y, and Chen Y. Reinforcement Learning in Multiple-UAV Networks: Deployment and Movement Design. *IEEE Trans Veh Technol*. 2019;68(8):8036–8049.

[25] Wang C, Wang J, Shen Y, et al. Autonomous Navigation of UAVs in Large-Scale Complex Environments: A Deep Reinforcement Learning Approach. *IEEE Trans Veh Technol*. 2019;68(3):2124–2136.

[26] Gu S, Holly E, Lillicrap T, et al. Deep Reinforcement Learning for Robotic Manipulation with Asynchronous Off-Policy Updates. In: *Proc. IEEE International Conf. Robot. Autom. (ICRA)*; 2017. p. 3389–3396.

[27] Cai Q, Filos-Ratsikas A, Tang P, et al. Reinforcement mechanism design for fraudulent behaviour in e-commerce. In: *Proc. Thirty-Second AAAI Conf. Artif. Intell.*; 2018.

[28] Mnih V, Kavukcuoglu K, Silver D, et al. *Playing Atari with Deep Reinforcement Learning*; 2013. Available from: https://arxiv.org/abs/1312.5602.

[29] Yu Y, Wang T, and Liew SC. Deep-Reinforcement Learning Multiple Access for Heterogeneous Wireless Networks. *IEEE J Select Areas Commun*. 2019;37(6):1277–1290.

[30] Zhao N, Liang YC, Niyato D, et al. Deep Reinforcement Learning for User Association and Resource Allocation in Heterogeneous Cellular Networks. *IEEE Trans Wireless Commun*. 2019;18(11):5141–5152.

[31] Yin S, Zhao S, Zhao Y, et al. Intelligent Trajectory Design in UAV-Aided Communications With Reinforcement Learning. *IEEE Trans Veh Technol*. 2019;68(8):8227–8231.

[32] Yang D, Wu Q, Zeng Y, et al. Energy Tradeoff in Ground-to-UAV Communication via Trajectory Design. *IEEE Trans Veh Technol*. 2018;67(7):6721–6726.

[33] Wang H, Wang J, Ding G, et al. Completion Time Minimization with Path Planning for Fixed-Wing UAV Communications. *IEEE Trans Wireless Commun*. 2019;18(7):3485–3499.

[34] Nguyen HT, Tuan HD, Duong TQ, *et al*. Joint D2D Assignment, Bandwidth and Power Allocation in Cognitive UAV-Enabled Networks. *IEEE Trans Cogn Commun Netw*. 2020;6(3):1084–1095.

[35] Liu L, Zhang S, and Zhang R. Multi-Beam UAV Communication in Cellular Uplink: Cooperative Interference Cancellation and Sum-Rate Maximization. *IEEE Trans Wireless Commun*. 2019;18(10):4679–4691.

[36] Xie L, Xu J, and Zhang R. Throughput Maximization for UAV-Enabled Wireless Powered Communication Networks. *IEEE Internet Things J*. 2019;6(2):1690–1703.

[37] Nguyen LD, Nguyen KK, Kortun A, *et al*. Real-Time Deployment and Resource Allocation for Distributed UAV Systems in Disaster Relief. In: *Proc. IEEE 20th International Workshop on Signal Processing Advances in Wireless Commun.* (SPAWC). Cannes, France; 2019. p. 1–5.

[38] Wu Q, Zeng Y, and Zhang R. Joint Trajectory and Communication Design for Multi-UAV Enabled Wireless Networks. *IEEE Trans Wireless Commun*. 2018;17(3):2109–2121.

[39] Zhan C, Zeng Y, and Zhang R. Energy-Efficient Data Collection in UAV Enabled Wireless Sensor Network. *IEEE Wireless Commun Lett*. 2018;7(3):328–331.

[40] Wang H, Ren G, Chen J, *et al*. Unmanned Aerial Vehicle-Aided Communications: Joint Transmit Power and Trajectory Optimization. *IEEE Wireless Commun Lett*. 2018;7(4):522–525.

[41] Wang Z, Liu R, Liu Q, *et al*. Energy-Efficient Data Collection and Device Positioning in UAV-Assisted IoT. *IEEE Internet Things J*. 2020;7(2):1122–1139.

[42] J. Li, H. Zhao, H. Wang *et al*. "Joint optimization on trajectory, altitude, velocity, and link scheduling for minimum mission time in UAV-aided data collection," *IEEE Internet Things J.*, vol. 7, no. 2, pp. 1464–1475, Feb. 2020.

[43] Samir M, Sharafeddine S, Assi CM, *et al*. UAV Trajectory Planning for Data Collection from Time-Constrained IoT Devices. *IEEE Trans Wireless Commun*. 2020;19(1):34–46.

[44] Hua M, Yang L, Wu Q, *et al*. 3D UAV Trajectory and Communication Design for Simultaneous Uplink and Downlink Transmission. *IEEE Trans Commun*. 2020;68(9):5908–5923.

[45] Zhan C and Zeng Y. Aerial–Ground Cost Tradeoff for Multi-UAV-Enabled Data Collection in Wireless Sensor Networks. *IEEE Trans Commun*. 2020;68(3):1937–1950.

[46] Zhang S and Zhang R. Radio Map-Based 3D Path Planning for Cellular-Connected UAV. *IEEE Trans Wireless Commun*. 2021;20(3):1975–1989.

[47] Zeng Y, Xu X, Jin S, *et al*. Simultaneous Navigation and Radio Mapping for Cellular-Connected UAV With Deep Reinforcement Learning. *IEEE Trans Wireless Commun*. 2021;20(7):4205–4220.

[48] Samir M, Assi C, Sharafeddine S, *et al*. Age of Information Aware Trajectory Planning of UAVs in Intelligent Transportation Systems: A Deep Learning Approach. *IEEE Trans Veh Technol*. 2020;69(11):12382–12395.

[49] Jaeger H. *The "Echo State" Approach to Analysing and Training Recurrent Neural Networks—With an Erratum Note*. GMD - German National Research Institute for Computer Science, Tech Rep. 2010;148(34):13.

[50] T. P. Lillicrap, J. J. Hunt, A. Pritzel *et al*. "Continuous control with deep reinforcement learning," in *Proc. 4th International Conf. on Learning Representations (ICLR)*, 2016.

[51] Wang Z, Schaul T, Hessel M, *et al*. *Dueling Network Architectures for Deep Reinforcement Learning*; 2015. Available from: https://arxiv.org/abs/1511.06581.

[52] Puterman ML. *Markov Decision Processes: Discrete Stochastic Dynamic Programming*. John Wiley & Sons, Inc.; 1994.

[53] Bertsekas DP. *Dynamic Programming and Optimal Control*. vol. 1. Athena Scientific Belmont, MA; 1995.

[54] M. Abadi, P. Barham, J. Chen *et al*. "Tensorflow: A system for large-scale machine learning," in *Proc. 12th USENIX Sym. Opr. Syst. Design and Imp. (OSDI 16)*, Nov. 2016, pp. 265–283.

[55] Kingma DP and Ba JL. *Adam: A Method for Stochastic Optimization*; 2014. Available from: https://arxiv.org/abs/1412.6980.

*Part IV*

**Deep reinforcement learning in reconfigurable intelligent surface-empowered 6G communications**

*Chapter 12*
# RIS-assisted 6G communications

Reconfigurable intelligent surfaces (RISs) are a cutting-edge technology poised to address key challenges in the development of the sixth-generation (6G) wireless networks. These challenges include minimising end-to-end communication latency and enhancing network reliability, surpassing the capabilities of the current 5G technology. Notably, the cost-effectiveness and energy-saving attributes of RISs have sparked a surge in research, focusing on their potential to optimise propagation environments and amplify received signal strength. This chapter provides a brief but complete introduction to RIS, starting by providing its main characteristics and benefits when applied in some specific scenarios and concluding by highlighting the main current challenges and identifying future research directions*.

## 12.1 Introduction

As already mentioned in Chapter 7, it is envisioned that with the deployment of next-generation wireless networks, i.e. 6G networks, it will be possible to provide connectivity to a large set of communication devices necessary for enabling the possibility to deliver a plethora of new innovative services with URLLC requirements. In other words, we will see a huge density of communication devices ($10^7$ devices/km$^2$) able to transfer/receive data with end-to-end latency less than $10\mu s$ and with 99.99999% communication reliability, providing then the possibility to deliver intelligent transportation, smart manufacturing, tactile internet and virtual/augmented reality (VR/AR) services. However, to achieve these demanding key performance indicators (KPIs) and enable the possibility to deploy such innovative services, notable improvements and advancements are necessary at different layers of the network architectures. In particular, great attention must be provided to the physical layer of the network by providing novel approaches for enhancing the signal propagation over the wireless medium, improving then the strength of received signals in challenging propagation environments like the one envisioned for the 6G network. Indeed, in addition to the huge diffusion of communication devices that will inevitably cause interference, the 6G networks are expected to adopt the THz frequency spectrum, which will add another challenge for signal propagation: high

---
*This chapter has been published partly in [1].

penetration losses. Then, it will be vital to design a highly efficient transceiver for signal transmissions and reception. This aspect has been partly covered with the advent of the 5G network, where the concepts of massive multiple-input multiple-output (MIMO) and hybrid analogue and digital beamforming technologies have been introduced [2,3]. However, designing high-frequency MIMO transceivers can be quite complicated. Recently, advances and improvements in the fabrication of electromagnetic (EM) metamaterials gave birth to the concept of reconfigurable intelligent surfaces (RISs), a new revolutionary physical layer technology which enables EM wave propagation control in a more energy-efficient and less complex way compared to the design of massive MIMO-oriented systems [4,5]. For these reasons, the adoption of RIS represents another key technology that is enabling the deployment of 6G-oriented systems. Under these perspectives, this chapter provides a brief but complete overview of the RIS technology and its role in wireless communications, as well as a discussion about its main related challenges that currently need to be addressed before this technology will be fully ready for its deployment.

## 12.2 RIS technology: a brief overview

The concept of RIS is based on the usage of metamaterial, a class of artificially engineered materials that exhibit reflective properties that cannot be found in nature. Indeed, these types of artificial materials are obtained by using micro or nano-scale structures, also known as meta-atoms, that are arranged into particular patterns with the purpose of achieving unusual properties. These properties can be used to obtain materials with negative refractive index as well as super- lensing properties, meaning that they would be able to either bend light around objects, making them invisible to certain wavelengths of light or create superlenses that can resolve details smaller than the wavelength of light. However, the properties of such metamaterials can also be extended in the frequency bands used to perform wireless communication transmissions. In these cases, the shape, size, and arrangement of the constituent micro/nanostructure can be modified to show different types of reflection, refraction and diffraction, depending on the wavelength of the incident EM wave.

The first appearance of metasurfaces can be tracked back to 2012, when, for the first time, they were introduced in conjunction with their respective generalised law for propagation [6]. That was an early stage of development within which metasurfaces were characterised by having a 3D structure, eventually miniaturised into a planar one, exhibiting the main issue that was not possible to obtain the desired response for the incident EM within real-time response requirements. Then, several research & development activities were carried out to address this issue, which led to the adoption of PIN diodes. Indeed, thanks to the adoption of these electronic devices and through the usage of a field-programmable gate array (FPGA), it was possible to obtain the response of the metasurface within real-time constraints [7]. The approach literally led to the realisation of RISs. These are basically physical objects having particular shapes (typically rectangular) and structures of meta-atoms, i.e. metasurfaces that can be externally programmed through appropriate signals to have the desired effect on the incident EM wave. This is possible through the employment of electronic phase transition elements like semiconductors or graphene as switches or alternatively by using tunable reactance/resistance elements between adjacent or within single meta-atoms. Such structures have represented the key idea that fostered the development of many system-level applications, such as space-time-coded digital communication systems and intelligent sensing systems for specific signal detection, either passive or active, which were previously difficult to achieve with traditional metamaterials [8].

## 12.3 RIS in next-generation wireless networks

In the context of wireless communications, the usage of RIS has been mainly investigated as a potential solution to improve the propagation environment. In other words, several studies have been carried out to understand how this technology can change and/or affect the considered wireless signals during their propagation period. As the main important results, these investigations showed how RISs would represent a very powerful technology that can enable various use cases by properly configuring amplitude, phase, frequency, and polarisation characteristics of the meta-atoms. More specifically, it has been found that it is possible to:

- Modify the amplitude characteristics of the RIS to mitigate unwanted signals such as interference. This involves accurately controlling the bias voltage of varactor diodes in each unit cell, resulting in a metamaterial with enhanced absorbance across a broad range of incident angles within the specified band.
- Adjust the phase characteristics to influence the reflection of the incident signal. This enables the creation of either single or multiple beams, with the reflected direction tailored to meet system requirements, such as extended coverage, jamming signals, and more.
- Adjust the frequency domain response to tune and potentially broaden the signal's frequency spectrum distribution. This capability stems from the distinctive non-linear response that can be achieved with RIS.
- Regulate the polarisation independently in each direction to modify the phase and amplitude responses of RIS elements to each incoming electromagnetic wave within the designated bandwidth.

While a RIS offers up to four degrees of freedom, the current predominant application involves leveraging its phase regulation capability to address various challenges in wireless communication. This makes it a potent tool for substantially enhancing communication system performance. The widely adopted use of phase-tuned RIS focuses on creating additional links to compensate for significant path loss and channel sparsity, thereby improving the effective connections between the base station and users. Moreover, RISs are employed to concentrate the reflected signal in specific directions, which is made possible through the joint optimisation of reflection coefficients and the phase matrix of the RIS. In other words, RIS-assisted communications can achieve heightened levels of spectral/energy efficiency and signal-to-noise ratio (SNR) for the end-to-end link, presenting a more cost-effective and lightweight alternative to deploying large-scale MIMO systems.

However, optimising phase-shift matrices in RIS-assisted communications requires a robust channel estimation of the surrounding environment. This task is not only more challenging than conventional scenarios but it is also anticipated to become increasingly demanding with the advent of higher frequencies in 6G and a growing number of users. Higher propagation losses and interference associated with these factors contribute to the heightened complexity of this optimisation process. Below, we list some typical RIS applications in various emerging 5G systems.

### 12.3.1 RIS-assisted multicell networks

Maximising the spectral efficiency (SE) of the networks represents an important task that is gaining even more attention with the deployment of the next generation of wireless networks. One approach consists of multiple BSs reusing the same frequency bands within the area of interest. However, as a side effect, this leads to inter-cell interference, which will inevitably penalise the performance of the system in terms of transmission capacity and connection reliability. More specifically, this will significantly impact cell-edge users in terms of suffering from a low signal-to-interference-plus noise ratio (SINR). In this case, using RIS has been shown to be beneficial in maximising the SINR of the users. Indeed, through numeric simulations, the authors in [9] illustrated how the usage of an RIS-aided system consisting of RISs placed at the cell boundary may double the achievable sum rate of the users covered by the cell.

### 12.3.2 RIS-aided non-orthogonal multiple access

Another way to improve SE that has also gained a lot of attention is the adoption of the non-orthogonal multiple access (NOMA) technique. Indeed, this approach enables the possibility of each orthogonal resource block being simultaneously shared by multiple users, which, compared with the traditional orthogonal multiple access (OMA), enhances the SE of the entire system. However, in some cases, the use of NOMA may not be the most preferable option. Indeed, in the presence of a multi-antenna transmitter, the usage of spatial division multiple access (SDMA) is preferable to NOMA when the channel vectors are orthogonal to each other, while NOMA represents the most suitable one when the channel vectors have the same directions. Hence, a crucial inquiry for expanding the applications of NOMA to improve SE involves determining whether it is possible to manipulate the directions of users' channel vectors, specifically aligning one user's channel with those of others. Under this perspective, it has been illustrated how RIS in the system allows for advantageous manipulation of the wireless channel vectors for all users, facilitating the alignment of one user's channel vector with that of another [10].

### 12.3.3 RIS for simultaneous wireless information and power transfer

Within the 6G vision, it is expected that a huge amount of IoT devices will be deployed. These are basically energy-constrained and need the adoption of energy-efficient solutions that can increase their operational duration as much as possible, especially for those deployed in remote areas. To this end, simultaneous wireless information and power transfer (SWIPT) presents a promising approach. In this technique, a BS with a constant power supply broadcasts wireless signals simultaneously to both information receivers (IRs) and energy receivers (ERs). A critical challenge in SWIPT systems is the distinct power supply requirements for ERs and IRs. Notably, ERs necessitate a received power significantly higher than that required by IRs. Given this requirement, ERs need to be positioned in closer proximity to the BS compared to IRs to harvest sufficient power, as signal attenuation limits the practical operational range of ERs. To address this challenge, it has been illustrated how the deployment of a RIS in the vicinity of the ERs, represents a very beneficial approach [11].

### 12.3.4 RIS-assisted mobile edge computing networks

The mobile edge computing (MEC) paradigm is expected to assist in the deployment of different time-sensitive and computation-intensive applications such as VR/AR services. Indeed, the main idea behind MEC is that this communication paradigm moves the computing of traffic and services from a centralised cloud to the edge of the network and closer to the customer, reducing then the communication latency. In addition to reduced service latency, this approach also permits IoT devices to perform computation-intensive tasks that cannot be accomplished locally due to power constraints by offloading them to the edge server. However, in specific scenarios where these devices are distant from the MEC node, they may experience a reduced data offloading rate attributed to significant path loss, resulting in prolonged offloading delays and then overall service delay. In this scenario, the utilisation of RISs has been explored as a potential solution to tackle this issue, demonstrating that the implementation of an RIS-assisted MEC paradigm has the potential to, at the very least, halve the delay time for cell-edge users [12].

### 12.3.5 RIS for physical layer security

Due to the inherent broadcast nature of wireless transmission, security threats such as jamming attacks and the potential leakage of secure information are prevalent in wireless links. Recently, physical layer security (PLS) techniques have garnered significant research attention. These techniques offer advantages by avoiding complex key exchange protocols and proving suitable for applications with low latency requirements. Efforts to maximise the secure communication link's rate have included the proposal of both artificial noise and multiple antennas. However, limitations persist in achieving a high-security rate, particularly when legitimate users and eavesdroppers have correlated channels or when eavesdroppers are in closer proximity to the BS than legitimate users. To this end, the usage of RIS has become a very good candidate to address this issue. In particular, it has been highlighted how the adoption of RIS results very helpful in increasing the secrecy rate of legitimate users in challenging conditions where the eavesdropping channel is stronger than the legitimate communication channel, and they are also highly correlated in space [13].

## 12.4 RIS-empowered UAV-assisted communications

In Chapter 7, it has been outlined how, thanks to their affordability, compact size, high mobility, and versatile deployment options, UAVs are emerging as a compelling solution for a variety of wireless communication applications. UAVs can serve as aerial communication nodes, contributing to the improvement of coverage, capacity, reliability, and energy efficiency in wireless networks. However, the effective integration of UAVs into wireless networks necessitates the resolution of several technical challenges and fundamental issues. To address some of the key challenges in UAV communications and enhance overall network performance in terms of spectrum and energy efficiency, RISs represent a very powerful solution to be incorporated into UAV networks. Indeed, the introduction of RISs is poised to play a pivotal role in elevating existing UAV networks by augmenting coverage and fortifying resilience in the face of environmental changes. Here, we will discuss some key advantages of merging these two technologies.

### *12.4.1 Enhancing air-to-ground (A2G) links*

Thanks to their high level of mobility, it is possible to use UAVs as aerial base stations or relays, which represents a very cost-effective solution for offloading macro-cells and providing energy-efficient connectivity. Indeed, UAVs can be developed as affordable aerial communication platforms to extend network coverage to remote and challenging-to-access regions or add extra capacity to existing networks. For example, in the case of large-scale natural disasters and emergencies that may damage terrestrial networks significantly, the adoption of UAV-assisted communication has emerged as a promising technology, providing a fast, flexible, and reliable solution to overcome network failures in such scenarios. In addition, their adoption in such types of scenarios plays a crucial role in supporting search-and-rescue operations and disaster response. Moreover, UAVs will result an ideal solution to overcome the challenges related to the huge diffusion of IoT devices expected to be connected. In such contexts, RISs can enhance air-to-ground (A2G) links between aerial platforms and ground users, maximising the area of UAV coverage and increasing transmit power efficiency. By integrating RISs into UAV-assisted wireless networks, virtual line-of-sight (LoS) links between the UAVs and ground users can be established through passive beamforming. This implementation results in a broader coverage area, more efficient transmissions, and reduced UAV mobility requirements.

### *12.4.2 RIS-equipped UAV communications*

In this context, UAVs will be outfitted with RISs to establish LoS links between BSs and users situated in coverage holes or blind spots. UAVs equipped with RISs can additionally facilitate communication between aerial platforms, taking advantage of the reliability offered by air-to-air (A2A) links. The future landscape of wireless networks is shaped by a burgeoning demand for high mobile data rates, minimised end-to-end latencies, and connectivity across a spectrum of emerging applications such as the IoT and massive machine-type communication. To meet these escalating demands, dense cell deployment and millimeter-wave communications have become imperative. However, dense networks encounter challenges concerning backhaul transmission and interference. Moreover, millimetre-wave bands are susceptible to high propagation loss and sensitivity to blockage, necessitating the use of dependable LoS links. In this context, UAVs equipped with RISs can dynamically support these technologies by establishing LoS connections precisely where they are required.

## 12.5 Challenges and research directions ahead

Based on the discussions carried out so far, it is evident how the adoption of RIS in next-generation networks represents an appealing and promising solution to address some of the upcoming challenges that currently are a bottleneck toward the deployment of innovative services. However, there are still some practical RIS-related challenges that need to be addressed before this technology becomes fully employable.

### *12.5.1 Channel state estimation techniques*

The full exploitation of RIS capabilities is based on the proper design of the so-called phase-shift matrix. This basically is used to control the response of the RIS surface on the incident EM, i.e. reflect the signal in the desired direction and/or reduce the noise effect. To do that, it is essential to perform an estimation of the channel coefficients between the entities involved in the RIS-aided communication. Currently, this estimation process presents several issues.

First of all, it is based on the usage of pilot signals, periodically transmitted from the base station to the users, which are directly proportional to the number of reflecting elements. Then, if, from one side, increasing the number of reflecting elements would lead to better system performances, on the other hand, it would potentially lead to high estimation overhead. Another aspect is related to the availability of angle or location information of users that can facilitate the coefficient estimation that sometimes is not available. Open challenges that still need to be addressed are (i) how to reduce this channel estimation overhead, and (ii) the design of low-complexity yet high-performance angle/location algorithms for RIS-aided networks.

Second, most of the existing contributions are based on the assumption of perfect channel coefficient estimation, which is something unrealistic in practice. Indeed, for the estimation of the cascaded channel, i.e. BS-to-RIS-to-UE channel, results necessary to first estimate the direct BS-to-UE channel by switching off the RIS, and then estimate the overall channel by switching on the RIS, obtaining the cascaded CSI as the difference between the direct BS-to-UE channel response from the overall channel response. Basically, the direct BS-to-UE channel cannot be perfectly estimated; this will cause contamination in terms of estimation imperfection in the cascaded channel, which should be taken into account.

Last but not least, assuming that is possible to have a perfect channel estimation, another problem is related to the availability of a limited number of quantised phase shifts that may be used to reduce hardware costs and consumption. This imposes an inevitable quantisation noise on the phase shifts of the RIS elements.

### 12.5.2 RIS deployment strategies

In the realm of RIS-assisted communications, it becomes evident that the system's performance is significantly influenced by the deployment strategy of the RIS. Indeed, considering the aforementioned discussion, one can easily notice how the identification of the optimal location for the RIS involves taking into account crucial factors, including user distribution, the specific service intended for the targeted area, and the prevention of interference with other pre-existing networks. An optimal deployment strategy for the RIS holds the potential to substantially diminish inadvertent interference among operators, along with minimising the channel state information (CSI) estimation time, especially when RIS with sensing capabilities are strategically deployed. This means that the availability of planning tools for RIS deployment will result in a very beneficial solution. So far, no work has proposed potential approaches to tackle this. However, the adoption of a digital twin (DT) paradigm represents a very promising approach for addressing some of these RIS-related issues. As an illustrative example, leveraging a DT of the relevant environment can aid in determining the optimal location of RISs, consequently decreasing the duration required for the channel estimation process.

### 12.5.3 Mobility management

Mobility management becomes intricate in RIS-aided wireless networks due to the rapid movements of users. Traditional approaches may fall short, considering the disconnection risk between BS and users due to their rapid mobility. The passive nature of RISs compounds this challenge, as they lack the capability to send pilot signals for real-time tracking of user movements. This limitation becomes especially pronounced when direct links between the BS and users are obstructed, necessitating the exploration of agile mobility management schemes. Consequently, tracking roaming users becomes a sophisticated task, demanding innovative strategies to ensure uninterrupted connectivity in the face of dynamic user scenarios. Novel strategies are essential to address these challenges. Leveraging advanced mobility management schemes capable of adapting to the rapid movements of users is crucial. Also, in this case, the adoption of a DT-based approach represents a powerful and attractive research direction.

### 12.5.4 The use of AI in RIS-assisted communications

The optimisation of phase-shift matrices in RISs is crucial for enhancing system performance. In practical scenarios, RISs are equipped with numerous reflecting elements, often in the hundreds. Existing contributions in phase-shift design predominantly rely on model-based optimisation methods characterised by a large number of iterations to approximate a near-optimal solution. This complexity arises from the non-convex nature of phase-shift constraints and the objective function. Consequently, these methods pose challenges for real-time applications due to their high computational demands. To mitigate this challenge, the adoption of AI-based approaches emerges as an attractive alternative capable of extracting system features without relying on a specific mathematical model. Once trained, this method enables the identification of optimal solutions through straightforward algebraic calculations. Moreover, the trained model exhibits robustness against imperfections in CSI and hardware impairments, enhancing its practical applicability.

## 12.6 Summary

Within this chapter, the reader has been introduced to the role of RIS as a key enabling technology for next-generation networks. In particular, the chapter started by recalling how the huge diffusion of IoT devices within the 6G networks will foster the diffusion of immersive services with URLLC requirements, as well as the need for improvements at the physical layer of the network worth particular attention. At this point, it has been outlined how RIS technology represents a valid candidate in this regard. Then, a brief introduction about the working principle of RIS has been provided by also illustrating some use cases where this technology would definitively play a decisive role in improving the system performance. The chapter concluded by pointing out the main research challenges that currently need to be addressed to make RIS a fully employable and exploitable technology. Some applications of DRL in solving some challenges in RIS-assisted UAV communications will be illustrated in the next chapters of this book.

## References

[1] Masaracchia A, Huynh DV, Alexandropoulos GC, *et al.* Toward the Metaverse Realization in 6G: Orchestration of RIS-Enabled Smart Wireless Environments via Digital Twins. *IEEE Internet Things Mag.* 2024;7(2):22–28.

[2] Lu L, Li GY, Swindlehurst AL, *et al.* An Overview of Massive MIMO: Benefits and Challenges. *IEEE J Select Topics Signal Process.* 2014;8(5):742–758.

[3] Ahmed I, Khammari H, Shahid A, *et al.* A Survey on Hybrid Beamforming Techniques in 5G: Architecture and System Model Perspectives. *IEEE Commun Surveys Tuts.* 2018;20(4):3060–3097.

[4] Calvanese Strinati E, Alexandropoulos GC, Wymeersch H, *et al.* Reconfigurable, Intelligent, and Sustainable Wireless Environments for 6G Smart Connectivity. *IEEE Commun Mag.* 2021;59(10):99–105.

[5] Alexandropoulos GC, Lerosey G, Debbah M, *et al.* Reconfigurable Intelligent Surfaces and Metamaterials: The Potential of Wave Propagation Control for 6G Wireless Communications. *IEEE ComSoc TCCN Newslett.* 2020;6(1):25–37.

[6] Yu N, Genevet P, Kats MA, et al. Light Propagation with Phase Discontinuities: Generalized Laws of Reflection and Refraction. *Science.* 2011;334(6054):333–337.

[7] Yang B, Cao X, Xu J, et al. Reconfigurable Intelligent Computational Surfaces: When Wave Propagation Control Meets Computing. *IEEE Wireless Commun.* 2023;30(3):120–128.

[8] Tang W, Chen MZ, Dai JY, et al. Wireless Communications with Programmable Metasurface: New Paradigms, Opportunities, and Challenges on Transceiver Design. *IEEE Wireless Commun.* 2020;27(2):180–187.

[9] Pan C, Ren H, Wang K, et al. Multicell MIMO Communications Relying on Intelligent Reflecting Surfaces. *IEEE Trans Wireless Commun.* 2020;19(8):5218–5233.

[10] Ding Z and Poor HV. A Simple Design of IRS-NOMA Transmission. *IEEE Commun Lett.* 2020;24(5):1119–1123.

[11] Pan C, Ren H, Wang K, et al. Intelligent Reflecting Surface Aided MIMO Broadcasting for Simultaneous Wireless Information and Power Transfer. *IEEE J Sel Areas Commun.* 2020;38(8):1719–1734.

[12] Bai T, Pan C, Deng Y, et al. Latency Minimization for Intelligent Reflecting Surface Aided Mobile Edge Computing. *IEEE J Sel Areas Commun.* 2020;38(11):2666–2682.

[13] Cui M, Zhang G, and Zhang R. Secure Wireless Communication via Intelligent Reflecting Surface. *IEEE Wireless Commun Lett.* 2019;8(5):1410–1414.

*Chapter 13*
# Real-time optimisation in RIS-assisted D2D communications

In this chapter, we propose a DRL-based approach aimed at solving the complex optimisation problem of the network's sum rate in RIS-assisted device-to-device (D2D) communication. In this context, the main aim of the RIS is to mitigate the interference present in the network and enhance the signal between the D2D pair. This is done by jointly optimising the transmit power at the D2D transmitter and the phase shift matrix at the RIS to maximise the network sum rate. In doing so, a Markov decision process for such a scenario is formulated, and then the equivalent optimisation problem is solved through the proximal policy optimisation for solving the maximisation game. Simulation results show impressive performance in terms of the achievable rate and processing time*.

## 13.1 Introduction

Device-to-device (D2D) communications play a critical role in 5G networks by allowing users to communicate directly without the involvement of base stations. It helps reduce the latency and improve the information transmission efficiency [2,3]. In [2], the D2D transmitters harvest energy through the simultaneous wireless information and power transfer protocol (SWIPT). Then, a game theory approach was proposed to solve the power allocation and power splitting at SWIPT with pricing strategies for maximising the network performance. In [3], the optimised power allocation was proposed to maximise the energy efficiency (EE) performance at the D2D-based vehicle-to-vehicle communications, by following a machine learning-based approach. Authors in [4] proposed a three-stage wireless energy harvesting protocol for a relay-assisted network in a cognitive spectrum sharing paradigm. For the considered network scenario and algorithm, they provided a closed-form expression for the outage probability. Subsequently, through computer simulations, they showed how the most relevant parameters, like the energy harvesting constraint, the interference power constraints on the primary user network, and an interference imposed by the primary user network on the secondary user cognitive network, impact the outage probability.

*This chapter has been published partly in [1].

Reconfigurable intelligent surface (RIS), referring to the technology of massive elements of flexible reflection capability controlled by an intelligent unit, has recently attracted great attention from the research community as an efficient means to expand wireless coverage. The RIS can manage the incoming signal by a controller, which allows to efficiently adapt the angle of passive reflection from the transmitters towards the receivers [5–8]. In [6], the RIS harvests energy from the access point (AP) and uses it to reflect the signal in two phases. The AP beamforming vector, the RIS's phase scheduling, and the passive beamforming were optimised to maximise the information rate. In [7], a channel estimation scheme for a multi-user multiple-input multiple-output system has been designed with the support of double RIS panels.

Some studies have investigated the efficiency of the RIS in assisting the D2D communications [9–13]. In [9] and [10], two sub-problems with fixed passive beamforming vector and fixed phase shift matrix were considered. To solve the power allocation optimisation with the fixed phase shift matrix, the authors in [9] used the gradient descent method, while the authors in [10] employed the Dinkelbach method. For the phase shift optimisation, a local search algorithm was proposed in [9] while fractional programming was utilised in [10]. However, these approaches assume a discrete phase shift and only reach a sub-optimal solution. Moreover, these works only consider perfect conditions, e.g. channel state information (CSI). In addition, these algorithms cause large delays due to high computational complexity.

Very recently, deep reinforcement learning (DRL) has been applied as an effective solution for solving complicated problems in wireless networks [14–18]. In [3], the DRL algorithm was used to choose the continuous transmit power level at the D2D transmitters for maximising the EE performance. In [15], discrete and continuous action spaces were considered for the beamforming vector and the RIS phase shift in multiple-input single-output communications. Then, two DRL algorithms were used to maximise the total throughput. In [16], a method based on the DRL was used for optimising the unmanned aerial vehicle (UAV)'s altitude and the RIS diagonal matrix to minimise the sum age-of-information. In [17], the authors used the DRL technique to maximise the signal-to-noise ratio.

Solving the joint optimisation of power allocation and RIS configuration remains a challenging problem. The traditional optimisation approaches mostly focus on solving sub-problems [9–13] or considering a discrete phase shift matrix at RIS [9,11] to reduce the complexity. In contrast, as already mentioned before, the adoption of DRL-based algorithms represents a very powerful and efficient approach for solving non-convex and complex problems. In this chapter, we propose a DRL algorithm for solving the joint power allocation and phase shift matrix optimisation in RIS-assisted D2D communications. First, we conceive a D2D communication system with the support of the RIS. The D2D channel is a combination of the direct link and the reflective link. In this context, the RIS is used to mitigate channel interference as well as to enhance information transmission. Second, we formulate a Markov decision process (MDP) [19] for the network throughput maximisation in the RIS-assisted D2D communications, in which the optimisation variables are the power at the D2D users and the phase shifts at the RIS. In this chapter, we characterise the continuous action space and propose an on-policy algorithm to search for an optimal policy for maximising the network sum rate. Therefore, we reduce the human intervention for designing the discrete variables, reduce neural networks' size, and train them better in centralised learning. Finally, we compare the efficiency of our proposed methods with other schemes in terms of the achievable network sum rate.

## 13.2 System model and problem formulation

We consider an RIS-assisted wireless network with $N$ pairs of D2D users distributed randomly and a RIS panel, as shown in Figure 13.1. Each pair of D2D users comprises a single-antenna D2D transmitter (D2D-Tx) and a single-antenna D2D receiver (D2D-Rx). A RIS panel with $K$ reflective elements is deployed to enhance the signal from the D2D-Tx to the associated D2D-Rx and mitigate interference from other D2D-Txs. The RIS with reflective elements maps the receiver's signal by the value of the phase shift matrix controlled by an intelligent unit. The received signal at the D2D-Rx is composed of a direct and reflective signal.

We denote the position of the $n$th D2D-Tx at time step $t$ as $X_n^t(\text{Tx}) = (x_n^t(\text{Tx}), y_n^t(\text{Tx})), n = 1, \ldots, N$ and that of the $\ell$th D2D-Rx as $X_\ell^t(\text{Rx}) = (x_\ell^t(\text{Rx}), y_\ell^t(\text{Rx})), \ell = 1, \ldots, N$. The RIS is fixed at the position $(x_{\text{RIS}}^t, y_{\text{RIS}}^t, z_{\text{RIS}}^t)$. The phase shift value of each element in the RIS belongs to $[0, 2\pi]$.

We denote the direct channel from the $n$th D2D-Tx to the $\ell$th D2D-Rx at time step $t$ by $h_{n\ell}^t$, and the reflective channel by $H_{n\ell}^t$. The phase shift matrix at the RIS at time step $t$ is defined by $\Phi^t = \text{diag}(\eta_1^t e^{j\theta_1^t}, \eta_2^t e^{j\theta_2^t}, \ldots, \eta_K^t e^{j\theta_K^t})$, where $\eta_k^t \in [0, 1]$ and $\theta_k^t \in [0, 2\pi]$ represent the reflection amplitude and the phase shift value, respectively; $j$ is the imaginary unit. In this chapter, we assume that the amplitudes of all elements are set to $\eta_k^t = 1$.

The distance between the $n$th D2D-Tx and the $\ell$th D2D-Rx at time step $t$ is defined as:

$$d_{n\ell}^t = \sqrt{(x_n^t(\text{Tx}) - x_\ell^t(\text{Rx}))^2 + (y_n^t(\text{Tx}) - y_\ell^t(\text{Rx}))^2}. \tag{13.1}$$

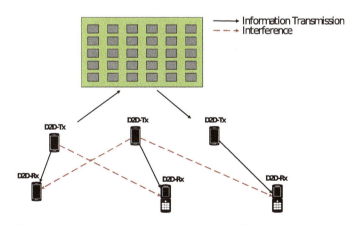

*Figure 13.1  System model of RIS-assisted D2D communications*

Similarly, the distance between the $n$th D2D-Tx and the RIS is $d_{n,RIS}^t$ and the distance between the RIS and the $\ell$th D2D-Rx is $d_{RIS,\ell}^t$ at time step $t$. The direct channel is formulated as:

$$h_{nm}^t = \sqrt{\beta_0 (d_{n\ell}^t)^{-\kappa_0}}, \tag{13.2}$$

where $\beta_0$ is the channel power gain at the reference distance $d_0 = 1$ m and $\kappa_0$ is the path-loss exponent in the line-of-sight (LoS) case. Here, we assume that the small-scale fading follows the Nakagami-$m$ distribution with $m$ as the fading severity parameter.

The channel gain between the $n$th D2D-Tx and the RIS is written as:

$$h_{n,RIS}^t = \sqrt{\beta_0 (d_{n,RIS}^t)^{-\kappa_1}} \left( \sqrt{\frac{\vartheta}{1+\vartheta}} \tilde{h}_{n,RIS}^{LoS} + \sqrt{\frac{1}{\vartheta+1}} \tilde{h}_{n,RIS}^{NLoS} \right), \tag{13.3}$$

where $\kappa_1$ is the path loss exponent, $\vartheta$ is the Rician factor; $\tilde{h}_{n,RIS}^{LoS}$ and $\tilde{h}_{n,RIS}^{NLoS}$ are the LoS and the non-line-of-sight (NLoS) components for the D2D-Tx and the RIS link, respectively. Specifically, the deterministic LoS component is defined as $\tilde{h}_{n,RIS}^{LoS} = [1, e^{-j\frac{2\pi}{\lambda} d \cos(\phi_{AoA}^t)}, \ldots, e^{-j\frac{2\pi}{\lambda} d(K-1) \cos(\phi_{AoA}^t)}]$, where $d$ and $\lambda$ are the RIS's element spacing and the carrier wavelength, respectively; $\cos(\phi_{AoA}^t)$ is the cosine of the angle of arrival(AoA). The NLoS component $\tilde{h}_{n,RIS}^{NLoS} \sim \mathcal{CN}(0, 1)$ follows i.i.d. complex Gaussian distribution with zero mean and unit variance. Similarly, the channel gain between the RIS and the $\ell$th D2D-Rx is $h_{RIS,\ell}$. The reflective channel via the RIS from the $n$th D2D-Tx towards the $\ell$th D2D-Rx at time step $t$ is described by $H_{n\ell}^t = h_{n,RIS}^t \Phi h_{RIS,\ell}^t$.

The received signal at the $n$th D2D-Rx at time step $t$ can be written as:

$$s_n^t = \left( h_{nn}^t + h_{n,RIS}^t \Phi h_{RIS,n}^t \right) \sqrt{p_n^t} u_n^t + \sum_{\ell \neq n}^{N} \left( h_{\ell n}^t + h_{\ell,RIS}^t \Phi h_{RIS,n}^t \right) \sqrt{p_\ell^t} u_\ell^t + \varpi, \tag{13.4}$$

where $p_n^t$ is the transmit power at the $n$th D2D-Tx at time step $t$, $u_n^t$ is the transmitted symbol from the $n$th D2D-Tx, and $\varpi \sim \mathcal{N}(0, \alpha^2)$ is the complex additive white Gaussian noise.

Accordingly, the received signal-to-interference-plus-noise ratio (SINR) at the $n$th D2D-Rx can be represented as:

$$\gamma_n^t = \frac{|h_{nn}^t + h_{n,RIS}^t \Phi h_{RIS,n}^t|^2 p_n^t}{\sum_{\ell \neq n, \ell \in N} |h_{\ell n}^t + h_{\ell,RIS}^t \Phi h_{RIS,n}^t|^2 p_\ell^t + \alpha^2}. \tag{13.5}$$

The achievable sum rate at the $n$th D2D pair during time step $t$ is defined as

$$R_n^t = B \log_2 (1 + \gamma_n^t), \tag{13.6}$$

where $B$ is the bandwidth.

In this chapter, we aim at optimising the power allocation of all $N$ pairs of D2D users $P = \{p_1, p_2, \ldots, p_N\}$ and the phase shift matrix $\Phi$ of the RIS to maximise the network sum rate while satisfying all the constraints. The considered network optimisation can be formulated as follows:

$$\max_{P,\Phi} R_{total}^t = \sum_{n=1}^{N} R_n^t$$
$$\text{s.t.} \quad 0 < p_n < P_{\max}, \forall n \in N \quad (13.7)$$
$$R_n^t \geq r_{\min}, \forall n \in N$$
$$\theta_k \in [0, 2\pi], \forall k \in K,$$

where $P_{\max}$ is the maximum transmit power at the D2D-Tx and the constraint $R_n^t \geq r_{\min}, \forall n \in N$ indicates the quality-of-service (QoS) of the D2D communications.

## 13.3 Joint optimisation of power allocation and phase shift matrix

Given the optimisation problem (13.7), we formulate the MDP with the agent, the state space $\mathscr{S}$, the action space $\mathscr{A}$, the transition probability $\mathscr{P}$, the reward function $\mathscr{R}$, and the discount factor $\varsigma$. Let us denote $\mathscr{P}_{ss'}(a)$ as the probability when the agent takes action $a^t \in \mathscr{A}$ at the state $s = s^t \in \mathscr{S}$ and transfers to the next state $s' = s^{t+1} \in \mathscr{S}$. We formulate the MDP game as follows:

- *State space*: The channel gain of the D2D users forms the state space as:

$$\mathscr{S} = \{|h_{11} + h_{1,RIS}^t \Phi h_{RIS,1}^t|^2, \ldots, |h_{1N}$$
$$+ h_{1,RIS}^t \Phi h_{RIS,N}^t|^2, \ldots, |h_{n\ell} + h_{n,RIS}^t \Phi h_{RIS,\ell}^t|^2,$$
$$\ldots, |h_{nN} + h_{n,RIS}^t \Phi h_{RIS,N}^t|^2, \ldots, |h_{N1} +$$
$$h_{n,RIS}^t \Phi h_{RIS,1}^t|^2, \ldots, |h_{NN} + h_{n,RIS}^t \Phi h_{RIS,N}^t|^2\}. \quad (13.8)$$

- *Action space*: The D2D-Txs adjust the transmit power and the RIS changes the phase shift for maximising the expected reward. Thus, the action space for the D2D users and the RIS is considered as follows:

$$\mathscr{A} = \{p_1, p_2, \ldots, p_N, \theta_1, \theta_2, \ldots, \theta_K\}. \quad (13.9)$$

- *Reward function*: The agent needs to find an optimal policy for maximising the reward. In our problem, our objective is to maximise the network sum rate; thus, the reward function is defined as:

$$\mathscr{R} = \sum_{n=1}^{N} B \log_2 \left(1 + \gamma_n^t\right). \quad (13.10)$$

In this chapter, we consider a centralised optimisation where the agent is considered as a central processor, for example, at a base station, on a powered D2D user or on the cloud. At the beginning of each time step, the agent transfers the action towards the D2D pairs and the RIS.

By following the MDP, the agent interacts with the environment and receives the response to achieve the best-expected reward. Particularly, the state of the agent at time step $t$ is $s^t \in \mathscr{S}$. The agent chooses and executes the action $a^t \in \mathscr{A}$ under the policy $\pi$. The environment responds with the reward $r^t \in \mathscr{R}$. After taking the action $a^t$, the agent moves to the new state $s'$ with probability $P_{ss'}(a)$. The interactions are iteratively executed, and the policy is updated to provide the optimal reward.

Next, we propose a DRL approach to search for an optimal policy for maximising the reward value in (13.10). The optimal policy can be obtained by modifying the estimation of the value function or directly by the objective. We use an on-policy algorithm for our work, namely proximal policy optimisation (PPO) with the clipping surrogate technique [20]. There are several advantages when designing the state space and action space in a continuous form. First, we can reduce human intervention while not needing to decide the number of discrete variables. Second, we can reduce the size of neural networks and train them better. For example, if we have $N$ D2D pairs with the power of each D2D pair being discretised into $J$ level and $K$ RIS elements with the phase shift of each element being divided into $K$ values, we need to define the output of the action-chosen neural network by $N \times J + K \times L$ in the centralised optimisation. In the meantime, we need only $N + K$ units for the output layer in the network when we use the continuous action space. Considering the probability ratio of the current policy and obtained policy $p_\theta^t = \frac{\pi(s,a;\theta)}{\pi(s,a;\theta_{old})}$, we need to find the optimal policy to maximise the total expected reward as follows:

$$\mathscr{L}(s,a;\theta) = \mathbb{E}\left[\frac{\pi(s,a;\theta)}{\pi(s,a;\theta_{old})} A^\pi(s,a)\right] = \mathbb{E}\left[p_\theta^t A^\pi(s,a)\right], \quad (13.11)$$

where $\mathbb{E}[\cdot]$ is the expectation operation and $A^\pi(s,a) = Q^\pi(s,a) - V^\pi(s)$ denotes the advantage function [21]; $V^\pi(s)$ denotes the state-value function while $Q^\pi(s,a)$ is the action-value function.

In the PPO method, we limit the current policy such that it does not go far from the obtained policy by using different techniques, e.g. the clipping technique and Kullback–Leiber [21]. We use the clipping surrogate method to prevent the excessive modification of the objective value, as follows:

$$\mathscr{L}^{\text{clip}}(s,a;\theta) = \mathbb{E}\left[\min\left(p_\theta^t A^\pi(s,a), \text{clip}(p_\theta^t, 1-\varepsilon, 1+\varepsilon) A^\pi(s,a)\right)\right], \quad (13.12)$$

where $\varepsilon$ is a hyperparameter.

Consider the positive value of the advantage $A^\pi(s,a)$ function and once $\pi(s,a;\theta) > (1+\varepsilon)\pi(s,a;\theta_{old})$, the term $(1+\varepsilon)$ takes action and the objective is limited by $(1+\varepsilon)A^\pi(s,a)$. We have:

$$\mathscr{L}^{\text{clip}}(s,a;\theta) = \min\left(\frac{\pi(s,a;\theta)}{\pi(s,a;\theta_{old})}, (1+\varepsilon)\right) A^\pi(s,a). \quad (13.13)$$

Meanwhile, when the advantage $A^\pi(s,a)$ is negative and $\pi(s,a;\theta) < (1-\varepsilon)\pi(s,a;\theta_{old})$, the term $(1-\varepsilon)$ puts a ceiling to the objective value and the objective is limited by $(1-\varepsilon)A^\pi(s,a)$. We have:

$$\mathscr{L}^{\text{clip}}(s,a;\theta) = \max\left(\frac{\pi(s,a;\theta)}{\pi(s,a;\theta_{old})}, (1-\varepsilon)\right) A^\pi(s,a). \quad (13.14)$$

Moreover, for the advantage function $A^\pi(s,a)$, we use [22]:

$$A^\pi(s,a) = r^t + \zeta V^\pi(s^{t+1}) - V^\pi(s^t), \tag{13.15}$$

where the state-value function $V^\pi(s)$ is obtained at the state $s$ under the policy $\pi$ as follows:

$$V^\pi(s) = \mathbb{E}\big[\mathcal{R}|s,\pi\big]. \tag{13.16}$$

To train the policy network, we store the transition into a mini-batch memory $D$ and then use stochastic policy gradient (SGD) method to maximise the objective. By denoting the policy parameter by $\theta$, it is updated as:

$$\theta^{i+1} = \arg\max \mathbb{E}\big[\mathcal{L}(s,a;\theta)\big]. \tag{13.17}$$

In this work, we use a policy search algorithm for an optimal policy $\pi^*$ with the policy parameter $\theta_\pi$. The PPO algorithm is an on-policy method; thus, we initialise a network for the policy $\pi$. After each interaction with the environment, the transition $(s^t, a^t, r^t, s')$ is stored in a buffer $D$. Then, the policy network is trained by the SGD with Adam optimiser [23] over $D$ samples. The policy parameters are updated by executing (13.17). Moreover, we use the advantage function to define the PPO objective as in (13.15). Thus, we define a network with the parameter $\phi_\theta$ to calculate the value function (13.16). The value network parameters $\phi_\theta$ are updated by mean-square error using the SGD algorithm as follows:

$$\phi_\theta^{i+1} = \arg\min \frac{1}{D}\sum^D \big(V^\pi(s) - r\big)^2. \tag{13.18}$$

The PPO algorithm for joint optimisation of the transmit power and the phase shift matrix in the RIS-aided D2D communications is presented in Algorithm 13.1, where $M$ denotes the maximum number of episodes and $T$ is the number of iterations during a period of time.

## 13.4 Simulation results

For numerical results, we use Tensorflow 1.13.1 [24]. The RIS is deployed at the centre $(0, 0, 0)$, while the D2D devices are randomly distributed within a circle of 100 m from the centre. The maximum distance between the D2D-Tx and the associated D2D-Rx is set to 10 m. We assume $d/\lambda = 1/2$, and set the learning rate for the PPO algorithm to 0.0001. For the neural networks, we initialise two hidden layers with 128 and 64 units, respectively. All other parameters are provided in Table 13.1. We consider the following algorithms in the numerical results.

- **The proposed algorithm**: We use the PPO algorithm with the clipping surrogate technique to solve the joint optimisation of the power allocation at the D2D user and the RIS's phase shift matrix.
- **Maximal power transmission (MPT)**: We apply the equal power allocation for the transmission of D2D-Tx, where each D2D-Tx transmits with maximal power $P_{\max}$. We use the PPO algorithm to optimise the RIS's phase shift matrix.

**Algorithm 13.1** Proposed approach based on the PPO algorithm for RIS-assisted D2D communications

1: Initialise the policy $\pi$ with the parameter $\theta_\pi$
2: Initialise other parameters
3: **for** episode = $1, \ldots, M$ **do**
4:     Receive initial observation state $s^0$
5:     **for** iteration = $1, \ldots, T$ **do**
6:         Obtain the action $a^t$ at state $s^t$ by following the current policy
7:         Execute the action $a^t$
8:         Receive the reward $r^t$ according to (13.10)
9:         Observe the new state $s^{t+1}$
10:        Update the state $s^t = s^{t+1}$
11:        Collect set of partial trajectories with $D$ transitions
12:        Estimate the advantage function according to (13.15)
13:     **end for**
14:     Update policy parameters using SGD with mini-batch $D$

$$\theta^{i+1} = \arg\max \frac{1}{D} \sum^D \mathscr{L}^{\text{clip}}(s, a; \theta^t) \quad (13.19)$$

15:     Update value network parameters $\phi_\theta$ using the SGD

$$\phi_\theta^{i+1} = \arg\min \frac{1}{D} \sum^D \left(V^\pi(s) - r\right)^2 \quad (13.20)$$

16: **end for**

*Table 13.1 Simulation parameters*

| Parameters | Value |
|---|---|
| Bandwidth ($W$) | 1 MHz |
| Path-loss parameters | $\kappa_0 = 2.5, \kappa_1 = 3.6$ |
| Channel power gain | $-30$ dB |
| Fading parameter | $\mu = 3$ |
| Rician factor | $\vartheta = 4$ |
| Noise power | $\alpha^2 = -110$ dBm |
| Clipping parameter | $\varepsilon = 0.2$ |
| Discount factor | $\zeta = 0.9$ |
| Max number of D2D pairs | 10 |
| Initial batch size | $K = 128$ |

- **Random phase shift matrix selection (RPS):** We optimise the power allocation at the D2D-Tx with random selection of the phase shift matrix $\Phi$.
- **Without RIS:** The D2D-Tx transmits information without the support of the RIS. We optimise the power allocation by using the PPO algorithm.
- **Vanilla policy gradient method (VPG)** [25]: We use neural networks for deploying a classical policy gradient method to optimise the power allocation of the D2D-Txs and the RIS's phase shift matrix.

First, we compare the achievable network sum rate provided by our proposed algorithm with that of other schemes. Figure 13.2 plots the sum rate versus different numbers of the RIS elements, $K$, where the number of D2D pairs is set to $N = 5$. As can be observed from this figure, the PPO algorithm-based technique outperforms other schemes and is followed by the MPT technique. The RPS, WithoutRIS, and VPG schemes show poorer performance in terms of the network sum rate. The achievable network sum rate using our proposed algorithm and MPT improves with increasing the number of RIS elements. The results show that with the monotonic increase in the value of $K$, the communication quality between the D2D-Tx and associated D2D-Rx is enhanced, while the interference from other D2D-Txs is suppressed.

Next, the performance of the previously mentioned four schemes is compared while varying the number of D2D pairs, $N$, in Figure 13.3. We set the number of RIS elements to $K = 50$ and take the average of over 500 episodes to obtain the results. Our proposed algorithm shows better performance, followed by MPT. With a higher number of D2D users, $N \geq 6$, the performance attained by the proposed algorithm still stables while it decreases significantly for the other schemes. The RPS and WithoutRIS models show the worse performance.

Further, we set $N = 5$, $K = 50$ and compare the performance results of the four schemes while changing the value of the threshold, $r_{\min}$, in Figure 13.4. When the value of $r_{\min}$ increases towards infinity, the number of D2D pairs that satisfy the QoS constraints decreases, and the sum rate of all schemes tends to 0. The proposed algorithm outperforms the other schemes for all values of $r_{\min}$. The gap between our algorithm and others increases following the increase in $r_{\min}$ when $r_{\min} \geq 15$ dB. The MPT algorithm exhibits the worst performance when $r_{\min} = 20$ dB. This suggests that the optimisation of power allocation is important for efficient D2D communications.

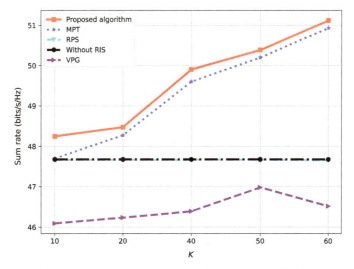

*Figure 13.2   Network sum rate versus the number of RIS elements, K*

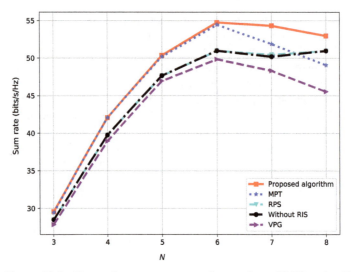

*Figure 13.3  Network sum rate versus the number of D2D pairs, N*

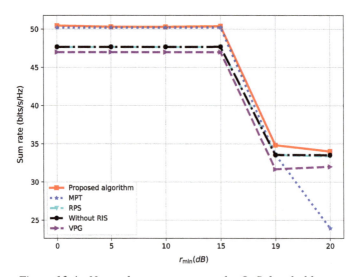

*Figure 13.4  Network sum rate versus the QoS threshold, $r_{\min}$.*

Next, we compare the total sum rate of the four schemes by setting different maximum transmission powers at the D2D-Tx, $P_{\max}$, in Figure 13.5, with $N = 5$, $K = 50$. As $P_{\max}$ varies from 200 mW to 400 mW, the performance of the five schemes increases in the same upward trend. The gap between our proposed algorithm and the other schemes increases with the increased value of $P_{\max}$ as we jointly optimise both power allocation at the D2D-Tx and the RIS's phase shift matrix. It is clear that the proposed algorithm is more effective for mitigating interference and providing better-quality communication.

*Real-time optimisation in RIS-assisted D2D communications* 213

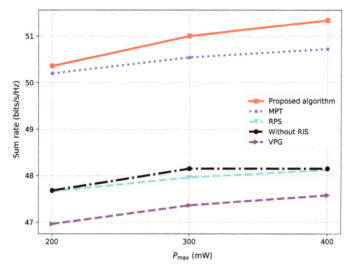

*Figure 13.5 Network sum rate versus the maximum transmit power, $P_{\max}$.*

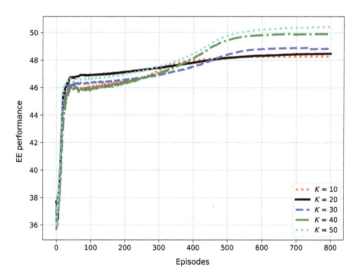

*Figure 13.6 The network sum rate while using the PPO algorithm*

Furthermore, we use neural networks to establish the DRL algorithm. Thus, after iterative interactions with the environment, the neural networks are trained to achieve an optimal solution. After training offline, the neural network can be deployed to the system for online execution. The online neural networks can determine the proper action for the RIS phase shift value and the D2D-Tx power allocation for maximising the network sum rate in real time.

In Figure 13.6, we compare the convergence speed of the PPO algorithm while varying the number of RIS elements, $K$. The PPO algorithm converges faster with the lower value of $K$. The slower convergence speed with the higher value of $K$ is mainly caused due to the higher number of optimisation variables.

## 13.5 Summary

This chapter has presented a DRL-based optimal resource allocation scheme for RIS-assisted D2D communications. More specifically, it has been illustrated how it is possible to use the PPO algorithm with the clipping surrogate technique for jointly optimising the transmit power of the D2D-Tx and the phase shift matrix of the RIS. Numerical results have shown a significant improvement in the achievable network sum-rate performance compared with the benchmark schemes, as well as the effectiveness of RIS in mitigating interference in the D2D communications when compared with other existing schemes.

## References

[1] Nguyen KK, Masaracchia A, and Yin C. Deep Reinforcement Learning for Intelligent Reflecting Surface-assisted D2D Communications. *EAI Endorsed Trans Ind Net Intell Syst.* 2023;10(1).

[2] Huang J, Xing CC, and Guizani M. Power Allocation for D2D Communications With SWIPT. *IEEE Trans Wireless Commun.* 2020;19(4):2308–2320.

[3] Nguyen KK, Duong TQ, Vien NA, et al. Distributed Deep Deterministic Policy Gradient for Power Allocation Control in D2D-Based V2V Communications. *IEEE Access.* 2019;7:164533–164543.

[4] Mousavifar SA, Liu Y, Leung C, et al. Wireless Energy Harvesting and Spectrum Sharing in Cognitive Radio. In: *Proc. IEEE 80th Vehicular Technology Conference (VTC2014-Fall)*, Vancouver, BC, Canada; 2014. p. 1–5.

[5] Yu H, Tuan HD, Nasir AA, et al. Joint Design of Reconfigurable Intelligent Surfaces and Transmit Beamforming Under Proper and Improper Gaussian Signaling. *IEEE J Select Areas Commun.* 2020;38(11):2589–2603.

[6] Zou Y, Gong S, Xu J, et al. Wireless Powered Intelligent Reflecting Surfaces for Enhancing Wireless Communications. *IEEE Trans Veh Technol.* 2020;69(10):12369–12373.

[7] Zheng B, You C, and Zhang R. Efficient Channel Estimation for Double-IRS Aided Multi-User MIMO System. *IEEE Trans Commun.* 2021;69(6):3818–3832.

[8] Nguyen KK, Khosravirad S, Costa DBD, et al. Reconfigurable Intelligent Surface-assisted Multi-UAV Networks: Efficient Resource Allocation with Deep Reinforcement Learning. *IEEE J Select Topics Signal Process.* 2021;1–1. Early Access.

[9] Che Y, Lai Y, Luo S, et al. UAV-Aided Information and Energy Transmissions for Cognitive and Sustainable 5G Networks. *IEEE Trans Wireless Commun.* 2021;20(3):1668–1683.

[10] Jia S, Yuan X, and Liang YC. Reconfigurable Intelligent Surfaces for Energy Efficiency in D2D Communication Network. *IEEE Wireless Commun Lett.* 2021;10(3):683–687.

[11] Pradhan C, Li A, Song L, *et al*. Reconfigurable Intelligent Surface (RIS)-Enhanced Two-Way OFDM Communications. *IEEE Trans Veh Technol.* 2020;69(12):16270–16275.

[12] Cao Y, Lv T, Ni W, *et al*. Sum-Rate Maximization for Multi-Reconfigurable Intelligent Surface-Assisted Device-to-Device Communications. *IEEE Trans Commun.* 2021;69(11):7283–7296.

[13] Yang G, Liao Y, Liang YC, *et al*. Reconfigurable Intelligent Surface Empowered Device-to-Device Communication Underlaying Cellular Networks. *IEEE Trans Commun.* 2021;69(11):7790–7805.

[14] Nguyen KK, Vien NA, Nguyen LD, *et al*. Real-Time Energy Harvesting Aided Scheduling in UAV-Assisted D2D Networks Relying on Deep Reinforcement Learning. *IEEE Access.* 2021;9:3638–3648.

[15] Huang C, Mo R, and Yuen C. Reconfigurable Intelligent Surface Assisted Multiuser MISO Systems Exploiting Deep Reinforcement Learning. *IEEE J Select Areas Commun.* 2020;38(8):1839–1850.

[16] Shokry M, Elhattab M, Assi C, *et al*. Optimizing Age of Information Through Aerial Reconfigurable Intelligent Surfaces: A Deep Reinforcement Learning Approach. *IEEE Trans Veh Technol.* 2021;70(4):3978–3983.

[17] Feng K, Wang Q, Li X, *et al*. Deep Reinforcement Learning Based Intelligent Reflecting Surface Optimization for MISO Communication Systems. *IEEE Wireless Commun Lett.* 2020;9(5):745–749.

[18] Nguyen KK, Duong TQ, Do-Duy T, *et al*. *3D UAV Trajectory and Data Collection Optimisation via Deep Reinforcement Learning*; 2021. Available from: https://arxiv.org/abs/2106.03129.

[19] Bertsekas DP. *Dynamic Programming and Optimal Control*. vol. 1. Athena Scientific Belmont, MA; 1995.

[20] Schulman J, Wolski F, Dhariwal P, *et al*. *Proximal Policy Optimization Algorithms*; 2017.

[21] Schulman J, Moritz P, Levine S, *et al*. High-Dimensional Continuous Control Using Generalized Advantage Estimation. In: *Proc. 4th International Conf. Learning Representations (ICLR)*; 2016.

[22] Mnih V, Badia AP, Mirza M, *et al*. Asynchronous Methods for Deep Reinforcement Learning. In: *Proc. Int. Conf. Mach. Learn. PMLR*; 2016. p. 1928–1937.

[23] Kingma DP and Ba JL. *Adam: A Method for Stochastic Optimization*; 2014. Available from: https://arxiv.org/abs/1412.6980.

[24] Abadi M, Barham P, Chen J, *et al*. Tensorflow: A System for Large-Scale Machine Learning. In: *Proc. 12th USENIX Sym. Opr. Syst. Design and Imp. (OSDI 16)*; 2016. p. 265–283.

[25] Sutton RS, McAllester D, Singh S, *et al*. Policy Gradient Methods for Reinforcement Learning with Function Approximation. *In: Adv. Neural Inf. Process. Syst.*; 2000. p. 1057–1063.

*Chapter 14*
# RIS-assisted UAV communications for IoT with wireless power transfer using deep reinforcement learning

Supplying energy while maintaining seamless connectivity for Internet-of-Things (IoT) devices is particularly important in next-generation networks. In this context, we show a possible way to implement a simultaneous wireless power transfer and information transmission scheme for IoT devices with the support of RIS-aided UAV communications. The approach consists of two successive phases. First, IoT devices harvest energy from the UAV through wireless power transfer. Subsequently, the UAV collects data from the IoT devices through information transmission. Two scenarios are considered: one with a UAV hovering at a specific position and the other scenario with a mobile UAV. However, in both cases, the main aim is to maximise the total network sum rate. In doing so, it will be illustrated how a correspondent Markov decision process can be formulated and how it is possible to use DRL for jointly optimising trajectory and the power allocation of the UAV, the energy harvesting scheduling of IoT devices, and the phase-shift matrix of the RIS. Numerical results illustrate the effectiveness of the UAV's flying path optimisation and the network throughput of our proposed techniques compared with other benchmark schemes. It is worth mentioning that the proposed scheme provides significant improvement in processing time and throughput performance, which makes it well suitable for practical IoT applications.[*]

## 14.1 Introduction

Unmanned aerial vehicles (UAVs) have recently drawn considerable attention thanks to their agile mobility and cost-effectiveness. UAVs have been used for geometry monitoring, disaster relief [2], emergency services, and wireless networks [3]. UAVs can be deployed at sporting events or in rescue missions to provide and enhance connectivity to users in wireless networks. UAVs are also used as data collectors that fly to remote areas to collect sensor data [4]. However, restrictions regarding flying time and on-board processing ability are bottlenecks that must be dealt with in unexpected environments and complicated missions. Very recently, UAVs have been

---

[*]This chapter has been published partly in [1].

used as an energy supplier to energy-constrained IoT devices due to the fact that the UAV can easily be recharged at the docking station and the energy of these IoT devices is comparably smaller than UAV's capacity [5–9].

Reconfigurable intelligent surface (RIS) has emerged as a promising technology for future wireless networks. The arrival signal at a RIS is reflected towards the receiver by the RIS's passive elements operated by a module controller. The received signal at the users is composed of elements from the direct channel and the reflective link. It helps to increase the signal quality and reduce the interference. The RIS is usually deployed in high locations, such as buildings, to reduce the cost of establishing a new station. However, optimising RIS performance is still challenging due to the large number of elements and the controller's processing ability.

One area in which UAVs can be useful is to support IoT applications. Not only can they provide communication coverages, but since many IoT devices are energy-limited, they can also be power sources for such devices through downlink power transfer. A downlink power transfer and uplink information transmission protocol can implemented in two phases: wireless power transfer (WPT) and wireless information transmission (WIT). In the first phase, the IoT devices harvest energy from a base station (BS) or from the UAV. The harvested energy is then used to transmit local information to receivers or return it to the UAV and the BS. In this way, the IoT devices can obtain the energy to establish and maintain communication with the BS and the UAV by using such a downlink power transfer and uplink information transmission protocol.

Machine learning is an effective tool for optimising the performance of large-scale networks. One of the approaches is deep reinforcement learning (DRL), which is a combination of reinforcement learning and neural networks. In wireless networks, DRL algorithms are used for maximising the network performance, reducing power consumption and improving the processing time for real-time applications [3,10,11]. DRL algorithms are powerful in wireless networks because the agents do not need pre-collected data for training. Rather, DRL agents interact with their environment and establish training samples for the responses in those interactions. The neural networks are trained by up-to-date state transitions to adjust their parameters for maximising a designated reward. Then, the trained networks are deployed for real-time prediction.

## *14.1.1 State of the art*

UAV-assisted wireless communications have been widely used to enhance network coverage as well as network performance [2,3,12]. In [2,12], the authors used the UAV to establish the network for disaster relief missions. A UAV can also serve as an energy source provider for device-to-device communications [3]. Recently, RIS technology has been introduced as a low-cost and easily installed technology to mitigate interference and direct transmitted signals towards their receivers [13–16]. In [14], the authors considered two-way communications assisted by a RIS. The reciprocal channel to maximise the signal-to-interference-plus-noise ratios (SINR) and the non-reciprocal channel with the target of maximisation of the minimum SINR were considered. The gamma approximation was used for the reciprocal channel, while the semi-definite programming relaxation and a greedy-iterative method were used for the non-reciprocal channel. In [15], an iterative algorithm with low computation complexity was proposed to solve the joint optimisation of the transmit beamforming vector and the phase shift of an RIS under proper and improper Gaussian signalling. In [16], the authors optimised the beamforming matrices at the BS and the reflective vector at the RIS to minimise the total transmit power at multiple-input single-output (MISO) non-orthogonal multiple access (NOMA) networks. An algorithm based on the second-order cone programming-alternating direction method of multipliers was proposed to reach an optimal local problem.

By utilising both advantages of the UAV and the RIS, the received signal at the ground users is strengthened while the power consumption is reduced, and the flying time of the UAV can be extended [17–20]. In [18], the UAV's trajectory and the RIS's passive beamforming vector were optimised to maximise the average rate in RIS-assisted UAV communications. The problem was derived into two sub-problems; then, a closed-form phase shift algorithm was introduced to find the local optimal reflective matrix and the successive convex approximation was used to find the sub-optimal trajectory solution. In [19], the UAV acts as a mobile relay, and the RIS was used to provide ultra-reliable and low-latency communications between ground transmitter and ground IoT devices. The UAV's position, the RIS phase shift and the blocklength were optimised to minimise the total decoding error rate by using a polytope-based method, namely Nelder–Mead simplex.

Along with the development of IoT devices is the increased power supply for each device. However, not all the nodes are equipped with fixed power providers or solar batteries. Thus, we need to find a solution to provide power to the nodes. The downlink power transfer and uplink information transmission protocol is one of the solutions to enable the IoT devices to harvest energy from source providers and switch to information transmission in the uplink phase on demand [21–25]. That helps reduce the power consumption as well as the cables and wires for providing power. In [21], the authors designed a time-switching protocol for a RIS with the energy harvesting phase to charge the RIS capacitor and the signal reflecting phase to assist the transmission from the access point (AP) to the receivers. The AP's transmit beamforming, the RIS's phase scheduling and the passive beamforming were optimised to maximise the information rate. The resultant two sub-problems were solved following the conventional semi-definite relaxation method and monotonic optimisation. In [22], the transmit precoding matrices of the BS and the RIS's passive phase shift matrix were optimised for maximising the weight sum rate of all information receivers in power transfer scenarios.

The demand for a technique that is flexible and adaptive to changes in the environment while satisfying real-life constraints is rising, and DRL algorithms are among the most potential methods to deal with these problems in wireless networks [3,4,10,11]. Recently, DRL algorithms are also used for RIS-assisted wireless networks and have shown promising results [26–30]. The power allocation and the phase shift optimisation were optimised for maximising the sum rate in [26]. In [27], an RIS-assisted UAV was deployed for serving ground users. The trajectory and phase shift optimisation relying on DRL to maximise the sum rate and fairness of all users was proposed. In [28], the authors used an RIS to assist the secure communications against eavesdroppers. The DRL algorithms were used to optimise the BS beamforming, and the RIS's reflecting beamforming was shown to improve the secrecy rate and the quality-of-service satisfaction probability. In [29], a deep deterministic policy gradient was proposed to obtain the optimal phase shift matrix at the RIS to maximise the received signal-to-noise ratio (SNR) in a MISO system. In [30], the joint optimisation of the power and the RIS's phase shift in a multi-UAV-assisted network is considered. In Table 14.1 and Table 14.2, we compare our work to the existing methods in the literature. While other works only consider a simple problem, our chapter introduces a comprehensive approach with the joint optimisation of power allocation, energy harvesting time scheduling, UAV's trajectory design and RIS's phase shift adjustment. Furthermore, we utilise the advantage of deep neural networks in RL algorithm for fast deployment and real-time decision-making.

## 14.1.2 Contributions

Inspired by the aforementioned discussion, in this chapter, we consider the IoT wireless networks with the support of a UAV and one RIS and employ the downlink power transfer and uplink information transmission protocol for maximising the total

Table 14.1 A comparison with existing literature (Part 1)

|  | [18] | [21] | [26] | [27] | [31] | [28] |
| --- | --- | --- | --- | --- | --- | --- |
| UAV's trajectory design | ✓ |  |  | ✓ | ✓ |  |
| Sum-rate maximisation | ✓ | ✓ | ✓ | ✓ | ✓ |  |
| Power allocation |  | ✓ | ✓ |  |  | ✓ |
| Energy harvesting time optimisation |  | ✓ |  |  |  |  |
| RIS phase shift optimisation | ✓ | ✓ | ✓ | ✓ |  | ✓ |
| Random users |  |  | ✓ | ✓ | ✓ | ✓ |
| Reinforcement learning |  |  |  | ✓ | ✓ | ✓ |
| Deep neural networks |  |  |  | ✓ |  | ✓ |

Table 14.2 A comparison with existing literature (Part 2)

|  | [29] | [32] | [33] | [34] | Our work |
| --- | --- | --- | --- | --- | --- |
| UAV's trajectory design |  |  | ✓ |  | ✓ |
| Sum-rate maximisation |  |  |  | ✓ | ✓ |
| Power allocation |  | ✓ | ✓ |  | ✓ |
| Energy harvesting time optimisation |  |  |  |  | ✓ |
| RIS phase shift optimisation | ✓ | ✓ | ✓ | ✓ | ✓ |
| Random users |  | ✓ | ✓ | ✓ | ✓ |
| Reinforcement learning | ✓ |  | ✓ |  | ✓ |
| Deep neural networks | ✓ |  | ✓ | ✓ | ✓ |

network's sum rate. In particular, we adopt the harvest-then-transmit protocol, which means the IoT devices use all the harvested energy in the first phase for transmitting during the remaining time. Then, two DRL algorithms are deployed to solve the problem in RIS-assisted UAV communications. In summary, our main contributions are as follows:

- To characterise the agility of the UAV in supporting the energy harvesting (EH) and information transmission of IoT devices, we consider two scenarios of the UAV. First, the UAV is hovering at the centre of the cluster and provides energy to the IoT devices. The RIS helps to alleviate the uplink interference when the IoT devices transmit their information to the UAV. Second, the UAV is deployed in an initial location and required to find a better location for communication. In each location of the UAV's flying trajectory, the UAV's power, the EH time scheduling and the RIS's phase shift matrix are optimised for maximising the network throughput performance.
- For the defined problem, we formulate a Markov decision process (MDP) [35] with the definition of the state space, action space, and the reward function. Then, we propose a method based on deep deterministic policy gradient (DDPG) and proximal policy optimisation algorithm (PPO) for solving the maximisation game. Through the trial-and-learn ability of the DRL algorithm, the UAV can collect the data and train the neural networks to adapt to the dynamic environment.
- Our results suggest that with the support of the RIS, a better connection is established, and the overall performance is significantly improved. Moreover, deep neural networks, after training, can be used to make real-time decisions.

## 14.1.3 Organisation

In the remaining chapter, we present the system model of the UAV-assisted wireless communication with the support of the RIS and problem formulation in Section 14.2. The MDP for maximising the network throughput problem is introduced for the hovering UAV scenario in Section 14.3. In Section 14.4, the DDPG algorithm is deployed for continuous control of the UAV's trajectory, EH time scheduling and the phase shift matrix of the RIS. The PPO technique with the clipping method is introduced in Section 14.5 for improving the network throughput. The simulation results are presented in Section 14.6 to illustrate the efficiency of our proposed methods compared with other baseline schemes. Some existing problems and potential future research topics for the RIS and the UAV in real-life applications are discussed in Section 14.7.

## 14.2 System model and problem formulation

As illustrated in Figure 13.1, we consider that the system includes one single-antenna UAV and $N$ ground IoT devices. Here, we assume that the IoT devices are located randomly in a specific area from the centre in which the UAV is the centroid. Although the IoT devices can be mobilised before making a connection with the UAV, for

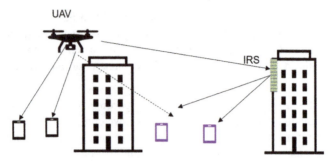

Figure 14.1 An illustration of the system model of UAV-assisted IoT wireless communications with the support of an RIS. © IEEE 2022. Reprinted with permission from [1].

the sake of simplicity, it is presumed that the devices are static during the power and information transmission. However, some practical scenarios exist where IoT devices are located in a crowded area with obstacles and objects. IoT devices suffer high attenuation and severe path loss in such a complex environment. In this case, the RIS is also installed at the wall of a tall building to enhance the communication quality by reflecting signals from the UAV to the IoT devices. Here, we deploy a RIS composed of $K$ elements to enhance the network performance.

In this chapter, the 3D coordinate of the UAV at the time step $t$ is $X_{\text{UAV}}^t = (x_{\text{UAV}}^t, y_{\text{UAV}}^t, z_{\text{UAV}}^t)$. We consider the fixed attitude of the UAV at $H_{\text{UAV}}$. The location of the $n$th IoT devices at time step $t$ is $X_n^t = (x_n^t, y_n^t)$ with $n = 1, \ldots, N$. The position of the RIS component $k \in K$ at time step $t$ is $(x_k^t, y_k^t, z_k^t)$.

We use the wireless downlink power transfer and uplink information transmission protocol to deploy the UAV and collect data. Before establishing the connection between the UAV and RIS, the UAV will interact with RIS via the control channel. Then, the wireless power and information transmission can be carried out through the physical channel. Particularly, we have two phases during the physical channel: wireless power transfer (WPT) and wireless information transmission (WIT). In the first phase, the downlink is activated to transfer energy to the IoT devices from the UAV during the time span $\tau \mathcal{T}$. Then, the WIT phase takes place when the IoT devices transmit information to the UAV in the uplink during $(1 - \tau)\mathcal{T}$. We normalise the length of time step to $\mathcal{T} = 1$ for convenience.

### 14.2.1 Channel model

We denote the channel gain between the UAV and the RIS, between the RIS and the $n$th IoT device, and the direct link from the UAV to $n$th IoT node at time step $t$ by $H^t \in \mathbb{C}^{1 \times K}$, $h_{\text{RIS},n}^t \in \mathbb{C}^{1 \times K}$, and $h_n^t$, respectively. The small-scale fading of the direct link from the UAV to the IoT devices is assumed to be Rayleigh fading due to the extensive scatters. The air-to-air channel is considered for the UAV and the RIS link, while the link from the RIS to the IoT devices can be modelled by the Rician fading channel. In practical applications, due to the acquisition delay and the feedback overhead incurred during the mobility of UAVs, obtaining a perfect CSI of the links between UAVs and RIS and IoT devices is a formidable challenge. In this chapter, we assume that the UAV and IoT devices are static, so the perfect CSI knowledge can be achieved [36].

The distance between UAV and the $k$th RIS in time step $t$ is given by:

$$d_k^t = \sqrt{(x_{UAV}^t - x_k^t)^2 + (y_{UAV}^t - y_k^t)^2 + (z_{UAV}^t - z_k^t)^2}. \tag{14.1}$$

Similarly, we denote the distance between the UAV and the $n$th IoT device and between the $k$th RIS element and the $n$th IoT node by $d_n^t$ and $d_{k,n}^t$, respectively.

The channel gain between the UAV and the $n$th IoT device is given by:

$$h_n^t = \sqrt{\beta_0(d_n^t)^{-\kappa_1}}\hat{h}, \tag{14.2}$$

where $\beta$ and $\kappa_1$ are the path loss at reference distance $1m$ and the path loss exponent for the UAV and the IoT devices link, respectively; $\hat{h}$ represents the small-scale fading modelled by complex Gaussian distribution with zero-mean and unit-variance $\mathcal{CN}(0, 1)$.

Similarly, the channel gain between the UAV and the RIS is an air-to-air channel dominated by the line-of-sight (LoS) links. Thus, the channel of the UAV-RIS link in time step $t$ is denoted as follows:

$$H^t = \sqrt{\beta_0(d_k^t)^{-\kappa_2}}[1, e^{-j\frac{2\pi}{\lambda}d\cos(\phi_{AoA}^t)}, \dots, \\ e^{-j\frac{2\pi}{\lambda}(K-1)d\cos(\phi_{AoA}^t)}]^T, \tag{14.3}$$

where the right term is the array signal from the UAV to the RIS, $\cos(\phi_{AoA}^t)$ is the cosine of the angle of arrival (AoA) from the UAV to RIS; $\kappa_2$, $d$, and $\lambda$ are the path loss exponent for the UAV and the RIS link, the element spacing and the carrier wavelength, respectively.

The channel gain between the RIS and the $n$th IoT device following the Rician fading is expressed as:

$$h_{RIS,n}^t = \sqrt{\beta_0(d_{k,n}^t)^{-\kappa_3}}\left(\sqrt{\frac{\beta_1}{1+\beta_1}}h_{RIS,n}^{LoS} + \sqrt{\frac{1}{\beta_1+1}}h_{RIS,n}^{NLoS}\right), \tag{14.4}$$

where the deterministic LoS component is denoted by $h_{RIS,n}^{LoS} = [1, e^{-j\frac{2\pi}{\lambda}d\cos(\phi_{AoD}^t)}, \dots, e^{-j\frac{2\pi}{\lambda}(K-1)d\cos(\phi_{AoD}^t)}]$ and the non-line-of-sight (NLoS) component is the Rayleigh fading that follows the complex Gaussian distribution with zero mean and unit variance; $\cos\phi_{AoD}$ is the angle of departure (AoD) from the RIS to IoT devices; $\beta_1$ is the Rician factor; and $\kappa_3$ is the path loss exponent for the RIS and IoT devices link.

## 14.2.2 Power transfer phase

The achievable signal at the $n$th IoT device is composed of the direct signal from the UAV and the reflected signal from the RIS at time step $t$ as:

$$y_{1n}^t = (h_n^t + H^t\Phi^t h_{RIS,n}^t)\sqrt{P_0}x + \varrho^2, \tag{14.5}$$

where $\varrho^2$ is the noise signal following the complex Gaussian distribution $\mathscr{CN}(0, \alpha^2)$, $x$ is the energy signal, and $P_0$ is the transmission power at the UAV; $\Phi^t = \text{diag}\,[\phi_1^t, \phi_2^t, \ldots, \phi_K^t]$ is the diagonal matrix at the RIS, where $\phi_k^t = e^{j\theta_k^t}, \forall k = 1, 2, \ldots, K$ and $\theta_k^t \in [0, 2\pi]$ denotes the phase shift of the $k$th element in the RIS at time step $t$.

In the WPT phase, the UAV transfers energy to the IoT devices during time span $\tau^t$ at time step $t$. Thus, the received power at the $n$th IoT device at time step $t$ is given by:

$$p_n^t = \tau^t \eta P_0 |h_n^t + H^t \Phi^t g_n^t|^2, \tag{14.6}$$

where $\eta$ is the power transfer efficiency. It is important to note that although we employ the linear EH model in this chapter, the non-linear counterpart should have been adopted in realistic scenarios.

### 14.2.3 Information transmission phase

We assume that the IoT devices do not have fixed energy sources and use all the harvested energy for the WIT phase. The signal received at the UAV from the $n$th IoT device is given by:

$$y_{2n}^t = (h_n^t + H^t \Phi^t h_{RIS,n}^t)\sqrt{p_n} u_n + \varrho^2, \tag{14.7}$$

where $u_n$ is the symbol signal from the $n$th IoT device to the UAV. The received SINR at the UAV from the transmission of the $n$th IoT device at time step $t$ can be formulated as follows:

$$\gamma_n^t = \frac{p_n^t |h_n^t + H^t \Phi^t g_n^t|^2}{\sum_{m \neq n}^N p_m^t |h_m^t + H^t \Phi^t g_m^t|^2 + \alpha^2}, \tag{14.8}$$

The sum rate from the IoT devices at time step $t$ is formulated as follows:

$$R_{\text{total}}^t = \sum_{n=1}^N (1 - \tau^t) B \log_2 (1 + \gamma_n^t), \tag{14.9}$$

where $B$ is the bandwidth.

Our objective is to maximise the achieved sum-rate performance by optimising the phase shift matrix $\Phi$ at the RIS, the UAV's trajectory $\Gamma$, power allocation $P_0$, and the EH time $\tau$ as:

$$\begin{aligned}
\max_{\tau, P_0, \Phi, \Gamma} & \sum_{n=1}^N (1 - \tau^t) B \log_2 (1 + \gamma_n^t), \\
\text{s.t.} \quad & 0 < \tau^t < 1, \\
& \theta_k^t \in [0, 2\pi], \forall k \in K, \\
& v^t \leq v_{\max}, \\
& 0 \leq P_0 \leq P_{\max}, \\
& X_{\text{UAV}}^t \in Z,
\end{aligned} \tag{14.10}$$

where $Z$ represents the flying area restricted in the vertical and horizontal dimensions; $P_{\max}$ is the maximum transmission power at the UAV; $v^t$ and $v_{\max}$ are the UAV's velocity at time step $t$ and the UAV's maximum flying velocity, respectively.

## 14.3 Hovering UAV for downlink power transfer and uplink information transmission in RIS-assisted UAV communications

Besides WPT, the UAV uses most of the energy for its movement. Thus, to extend the operating time, the UAV is considered to hover at a fixed position at the centre of the cluster. First, we present the DRL algorithm's mathematical background and then apply the DRL algorithm to solving the sum-rate maximisation problem in RIS-assisted UAV communications.

### 14.3.1 Preliminaries

To search for optimal policy in the DRL algorithm, we have two main directions: by directly searching the policy parameters and indirectly through the value function. In the value function approach, we estimate the value in a given state to adjust the policy parameter. The state-value function $V$ is defined by following the policy $\pi$ at state $s \in \mathcal{S}$ as follows:

$$V^\pi = \mathbb{E}\big[\mathcal{R}|s, \pi\big], \qquad (14.11)$$

where $\mathbb{E}$ is the expectation operation; $\mathcal{S}$ and $\mathcal{R}$ are the state space and the reward function, respectively.

The optimal policy $\pi^*$ corresponds to the optimal state-value function $V^*(s)$ as:

$$V^*(s) = \max_\pi V^\pi(s), s \in \mathcal{S}. \qquad (14.12)$$

The agent at state $s$ chooses the action $a$ from the action space $\mathcal{A}$ to maximise the cumulative reward following the optimal policy $\pi^*$ by making use of the Bellman equation [35]:

$$V^*(s) = \max_{a \in \mathcal{A}} \left\{ \mathbb{E}\big[r(s,a)\big] + \zeta \sum_{s' \in \mathcal{S}} \mathcal{P}_{ss'}(a) V^*(s') \right\}, \qquad (14.13)$$

where $\mathcal{P}_{ss'}(a)$ is the transition probability with $s = s^t$, $s' = s^{t+1} \in \mathcal{S}$, and $a \in \mathcal{A}$; $0 \leq \zeta \leq 1$ is the discount factor.

The state-action value $Q$ is obtained when the agent at state $s$ takes action $a \in \mathcal{A}$ following the policy $\pi$ as follows:

$$Q^\pi(s,a) = \mathbb{E}\big[r(s,a)\big] + \zeta \sum_{s' \in \mathcal{S}} \mathcal{P}_{ss'}(a) V(s'). \qquad (14.14)$$

From (14.13) and (14.14), we have:

$$V^*(s) = \max_{a \in \mathcal{A}} Q^*(s,a), \qquad (14.15)$$

where $Q^*(s,a)$ is the action-value function with optimal policy $\pi^*$.

Instead of finding the optimal policy $\pi^*$ through the value function, we directly find the policy by adjusting the policy parameter in the policy search method. The policy gradient is one of the most popular methods due to its efficient sampling. Our objective is to find the optimal policy $\pi^*$ to maximise the reward, which is defined by:

$$J(\theta_\pi) = \sum_{s \in \mathcal{S}} d^\pi(s) \sum_{a \in \mathcal{A}} \pi(a|s) r(s, a), \qquad (14.16)$$

where $\theta_p i$ and $d^\pi(s)$ are the policy parameter and the stationary distribution of the Markov chain, respectively. We adjust the policy parameter $\theta_p i$ by using the gradient descent relying on:

$$\begin{aligned}\nabla \theta_\pi &= \sum_{s \in \mathcal{S}} d^\pi(s) \sum_{a \in \mathcal{A}} \nabla_{\theta_\pi} \pi(a|s) Q^\pi(s, a) \\ &= \mathbb{E}\left[\nabla_{\theta_\pi} \ln \pi(s, a) Q^\pi(s, a)\right].\end{aligned} \qquad (14.17)$$

One of the most popular algorithms based on policy search is the REINFORCE algorithm, in which we adjust the policy parameter $\theta_p i$ by using the Monte-Carlo methods and episode samples learning. The optimal policy can be obtained by:

$$\theta_\pi^* = \underset{\theta_\pi}{\operatorname{argmax}} \mathbb{E}\Big[\sum_{a \in \mathcal{A}} \pi(a|s; \theta_\pi) r(s, a)\Big]. \qquad (14.18)$$

We use the gradient descent method to adjust the policy parameters $\theta_\pi$ as:

$$\theta_\pi \leftarrow \theta_\pi - \iota \nabla \theta_\pi, \qquad (14.19)$$

where $0 \leq \iota \leq 1$ is the step-size parameter, and the gradient is defined as:

$$\nabla \theta_\pi = \mathbb{E}\left[\nabla_{\theta_\pi} \ln \pi(a|s; \theta_\pi) r(s, a)\right]. \qquad (14.20)$$

At state $s$ we can choose the best action $a^*$ by the maximum probability as:

$$a^* = \underset{a \in \mathcal{A}}{\operatorname{argmax}} \pi(a|s; \theta_\pi). \qquad (14.21)$$

### 14.3.2 DDPG method

The DDPG algorithm is a hybrid model composed of the value function and policy search methods. Thus, the DDPG algorithm is suitable for large-scale action and state spaces. Based on the current policy, the actor function $\mu(s; \theta_\mu)$ maps the states to a specific action with $\theta_\mu$ being the actor-network parameters, while the critic function $Q(s, a)$ evaluates the quality of the action taken. In the DDPG algorithm, we use *experience replay buffer* and *target network* techniques to improve the convergence speed and avoid excessive calculations.

The agent iteratively interacts with the environment by executing the action $a^t$ and receiving the response with instant reward $r^t$ and the next state $s^{t+1}$. The tuple of $(s^t, a^t, r^t, s^{t+1})$ is then stored in a replay buffer $D$ for training the actor and critic network. The buffer $D$ is updated by adding new samples and discarding the oldest ones due to its finite-size setting. After achieving enough samples, the agent takes a batch $G$ of transitions to train the network. Particularly, we train the actor and critic network using stochastic gradient descent (SGD) over mini-batch $G$ samples.

Let us denote the parameters of the critic network and the target critic network by $\theta_q$ and $\theta_{q'}$, respectively. The critic network is updated by minimising the following quantity:

$$L = \frac{1}{G} \sum_{i}^{G} \left( y^i - Q(s^i, a^i; \theta_q) \right)^2, \tag{14.22}$$

with

$$y^i = r^i(s^i, a^i) + \zeta Q'(s^{i+1}, a^{i+1}; \theta_{q'})|_{a^{i+1} = \mu'(s^{i+1}; \theta_{\mu'})}. \tag{14.23}$$

where the action at time step $(i+1)$ can be obtained by running the target actor network $\mu'$ with the state $s^{i+1}$; $\theta_{\mu'}$ denotes the parameters of the target actor network and $\zeta$ is the discounting factor.

The actor-network parameters are updated by:

$$\nabla_{\theta_\mu} J \approx \frac{1}{G} \sum_{i}^{G} \nabla_{a^i} Q(s^i, a^i; \theta_q)|_{a^i = \mu(s^i)} \nabla_{\theta_\mu} \mu(s^i; \theta_\mu). \tag{14.24}$$

Moreover, we duplicate the actor-network and the critic network after a number of episodes to create a target actor and a target critic network. It helps reduce the excessive calculations by using only one network to estimate the target value. The target actor network parameters $\theta_q$ and the target critic network parameter $\theta_{\mu'}$ are updated by using soft target updates associated with $\varkappa \ll 1$, as:

$$\theta_{q'} \leftarrow \varkappa \theta_q + (1 - \varkappa)\theta_{q'}, \tag{14.25}$$

$$\theta_{\mu'} \leftarrow \varkappa \theta_\mu + (1 - \varkappa)\theta_{\mu'}. \tag{14.26}$$

For *explorations* and *exploitations* purpose, we add a noise process of $\mathcal{N}(0, 1)$ as follows [37]:

$$\mu'(s^t) = \mu(s^t; \theta_\mu^t) + \psi \mathcal{N}(0, 1), \tag{14.27}$$

where $\psi$ is a hyper-parameter.

### 14.3.3 Game solving

We formulate the MDP [35] by a 4-tuple $<\mathcal{S}, \mathcal{A}, \mathcal{P}, \mathcal{R}>$ for the hovering scenario. Then, we formulate a game to solve the problem in (14.10).

- *Agent*: The UAV's centralised processor will act as an agent. The agent interacts with the environment to find an optimal policy $\pi^*$ for maximising the total sum rate. After training, the action-making schemes will be deployed to the UAV to predict the proper EH time scheduling $\tau$, and the RIS can choose the phase shift matrix $\Phi$.
- *State space*: The channel is composed of both direct link and the reflective channel. Thus, we define the state space as:

$$\mathcal{S} = \{h_1 + H\Phi g_1, h_2 + H\Phi g_2, \ldots, h_N + H\Phi g_N\}, \tag{14.28}$$

The UAV has the state $s^t = \{h_1^t + H^t\Phi^t g_1^t, h_2^t + H^t\Phi^t g_2^t, \ldots, h_N^t + H^t\Phi^t g_N^t\}$ at time step $t$.

- *Action space*: The UAV hovers at a fixed position; thus, we optimise the EH time $\tau$, the transmission power $P_0$, and the RIS's phase shift $\Phi$. The action space is defined as:

$$\mathscr{A} = \{\tau, P_0, \theta_1, \theta_2, \ldots, \theta_K\}. \tag{14.29}$$

At the state $s^t$, the UAV takes the action $a^t = \{\tau^t, P_0^t, \theta_1^t, \theta_2^t, \ldots, \theta_K^t\}$ and moves to the next state $s' = s^{t+1}$.

- *Reward function*: The UAV interacts with the environment to find the maximum obtained reward. In our work, we formulate the reward function to obtain the maximum total sum-rate performance as:

$$\mathscr{R} = \sum_{n=1}^{N} (1 - \tau^t) B \log_2(1 + \gamma_n^t). \tag{14.30}$$

The UAV is hovering at $X_{\text{UAV}}$ and chooses the action $a^t$ based on the achieved channel state information (CSI). Then, the UAV, using the power $P_0$, transfers the energy during $\tau$ to the IoT devices, and the RIS controller adjusts the phase shift for each element. During the remaining time $(1 - \tau)$, the RIS will not change the phase shift while the IoT devices transmit information in the uplink to the UAV. It is challenging since the RIS plays a crucial role in mitigating the interferences. Thus, we need to find an intelligent scheme for the RIS to maximise the network performance in the downlink power transfer and uplink information transmission protocol. We propose a DRL, namely the DDPG algorithm, to find an optimal policy for the UAV and the RIS.

At time step $t$, the agent has the channel information (14.28) and uses the actor-network to choose the action $a^t$ (14.29). By executing the action $a^t$, the agent receives the response following the reward function (14.30) from the environment. The critic action takes part to justify the efficiency of the chosen action $a^t$. After storing enough samples in buffer $D$, the agent trains the networks over a mini-batch $G$ by using the SGD with Adam optimiser [38].

This section assumes that the UAV is hovering at a fixed position to reduce the flying energy consumption. It is a trade-off game with the energy and total achievable sum rate. In the next section, we propose a joint optimisation of trajectory, power allocation, EH time and the phase shift to maximise the network throughput in a short operation time.

## 14.4 Joint trajectory, transmission power, energy harvesting time scheduling, and the RIS phase shift optimisation using deep reinforcement learning

Given the short flying time of the UAV, to maximise the total achievable sum rate, we propose a joint optimisation scheme between the UAV's trajectory, transmission power, EH time scheduling of IoT, and the RIS's phase shift. We define the state space and the reward function as in Section 14.3. We modify the action space as follows:

$$\mathscr{A} = \{v, \varsigma, \tau, P_0, \theta_1, \theta_2, \ldots, \theta_K\}. \tag{14.31}$$

At the state $s^t$, the UAV takes the action $a^t = \{v^t, \varsigma^t, \tau^t, P_0^t, \theta_1^t, \theta_2^t, \ldots, \theta_K^t\}$ and moves to the next state $s' = s^{t+1}$. Particularly, the position of the UAV at time step $(t+1)$ is represented as follows:

$$X_{\text{UAV}}^{t+1} = \begin{cases} x_{\text{UAV}}^{t+1} = x_{\text{UAV}}^t + v^t \cos \varsigma^t + \Delta x^{t+1} \\ y_{\text{UAV}}^{t+1} = y_{\text{UAV}}^t + v^t \sin \varsigma^t + \Delta y^{t+1} \\ H_{\text{UAV}}^{t+1} = H_{\text{UAV}}^t + \Delta H^{t+1}, \end{cases} \tag{14.32}$$

where $\Delta x^{t+1}, \Delta y^{t+1}$, and $\Delta H^{t+1}$ are the environmental noise on the UAV at time step $(t+1)$. The UAV is flying from the position $X_{\text{UAV}}^t$ to $X_{\text{UAV}}^{t+1}$ but still needs to satisfy the flying zone constraint $X_{\text{UAV}} \in Z$. Moreover, the velocity of the UAV is set to satisfy the requirement $v \leq v_{\max}$ and the flying angle is set to satisfy a constraint, $\varsigma \in [0, 2\pi]$.

Our objective is to find the optimal policy $\pi^*$ for maximising the expected reward $\mathscr{R}$. The agent has the local knowledge and interacts with the environment to receive the reward. Based on the received reward, the agent adjusts the policy $\pi$ and executes a new action at a new state. The agent can find a better policy with a better reward by the iterative interactions. After each execution of the action, the UAV will move to a new position and receive responses from the environment. By interacting iteratively with the environment, the agent can choose the proper velocity and the flying direction for the UAV in each time step based on the achieved CSI. Simultaneously, the EH scheduling $\tau$, the power allocation $P_0$, and the phase shift matrix are also optimised for maximising network performance. Here, $M$ and $T$ are the number of the maximum episodes and time steps, respectively. The details of our DDPG algorithm-based technique for joint trajectory design, power allocation, EH time, and phase shift matrix optimisation in RIS-assisted UAV communications are presented in Algorithm 14.1.

## 14.5 Proximal policy optimisation technique for joint trajectory, power allocation, energy harvesting time, and the phase shift optimisation

For the continuous state and action space as in our problem, we propose an on-policy algorithm, namely the PPO algorithm, for the joint optimisation of trajectory, power allocation, EH time and the phase shift of the RIS. Instead of using a critic network based on value function in the Algorithm 14.1, we search directly for the optimal policy $\pi^*$ in the PPO algorithm. We define the policy by $\pi$ with the parameter $\theta_\pi$. Here, we train the policy and adjust the parameter to find an optimal policy $\pi^*$ by running

**Algorithm 14.1** Deep deterministic policy gradient algorithm for joint trajectory design, transmission power, EH time, and phase shift optimisation in RIS-assisted UAV communications

1: Initialise the actor network $\mu(s; \theta_\mu)$, target actor network $\mu'$ and the critic network $Q(s, a; \theta_q)$, the target critic networks $Q'$.
2: Initialise replay memory pool $\mathcal{D}$
3: **for** episode = $1, \ldots, M$ **do**
4:     Initialise an action exploration process $\mathcal{N}$
5:     Receive initial observation state $s^0$
6:     **for** iteration = $1, \ldots, T$ **do**
7:         Find the action $a^t$ for the state $s^t$
8:         Execute the action $a^t$
9:         Update the reward $r^t$ according to (14.30)
10:        Observe the new state $s^{t+1}$
11:        Store transition $(s^t, a^t, r^t, s^{t+1})$ into replay buffer $\mathcal{D}$
12:        Sample randomly a mini-batch of $G$ transitions $(s^i, a^i, r^i, s^{i+1})$ from $\mathcal{D}$
13:        Update critic parameter by SGD using the loss (14.22)
14:        Update the actor policy parameter (14.24)
15:        Update the target networks as in (14.25) and (14.26)
16:        Update the state $s_i^t = s_i^{t+1}$
17:     **end for**
18: **end for**

---

the SGD over a mini-batch of $G$ transitions $(s^i, a^i, r^i, s^{i+1})$. The policy parameters are updated for optimising the objective function as follows:

$$\theta_\pi^{i+1} = \underset{\theta_\pi}{\arg\max} \frac{1}{G} \sum_i^G \nabla_{a^i} \mathcal{L}(s^i, a^i; \theta_\pi). \tag{14.33}$$

In the PPO algorithm, the agent interacts with the environment to find the optimal policy $\pi^*$ with the parameter $\theta_{\pi^*}$ that maximises the reward as:

$$\mathcal{L}(s, a; \theta_\pi) = \mathbb{E}\left[p_\theta^t A^\pi(s, a)\right], \tag{14.34}$$

where $p_\theta^t = \frac{\pi(s,a;\theta_\pi)}{\pi(s,a;\theta_{old})}$ is the probability ratio of the current policy and previous policy; $A^\pi(s, a)$ is the advantage function [39].

Here, if we use only one network for the policy, excessive modification occurs during the training stage. Thus, we use the clipping surrogate method as follows [40]:

$$\mathcal{L}^{\text{clip}}(s, a; \theta_\pi) = \mathbb{E}\left[\min\left(p_\theta^t A^\pi(s, a), \text{clip}(p_\theta^t, 1 - \varepsilon, 1 + \varepsilon) A^\pi(s, a)\right)\right], \tag{14.35}$$

where $\varepsilon$ is a small constant. In this chapter, the advantage function $A^\pi(s, a)$ [41] is formulated as follows:

$$A^\pi(s, a) = r^t + \zeta V^\pi(s^{t+1}) - V^\pi(s^t). \tag{14.36}$$

*RIS-assisted UAV communications for IoT with WPT using DRL* 231

---

**Algorithm 14.2** Our proposed approach based on the PPO algorithm for the RIS-assisted UAV communications

---
1: Initialise the policy $\pi$ with the parameter $\theta_\pi$
2: Initialise the penalty method parameters $\varepsilon$
3: **for** episode = $1, \ldots, M$ **do**
4:     Receive initial observation state $s^0$
5:     **for** iteration = $1, \ldots, T$ **do**
6:         Find the action $a^t$ based on the current state $s^t$ by following the current policy
7:         Execute the action $a^t$
8:         Update the reward $r^t$ according to (14.30)
9:         Observe the new state $s^{t+1}$
10:        Update the state $s_i^t = s_i^{t+1}$
11:        Collect a set of partial trajectories with $G$ transitions
12:        Estimate the advantage function as (14.36)
13:     **end for**
14:     Update policy parameters using SGD with a mini-batch $B$ of the collected samples

$$\theta^{i+1} = \underset{\theta_\pi}{\mathrm{argmax}} \frac{1}{G} \sum^{G} \mathscr{L}^{\mathrm{clip}}(s, a; \theta_\pi) \tag{14.37}$$

15: **end for**

---

The policy is then trained by a mini-batch $B$ and the parameters are updated by:

$$\theta^{i+1} = \underset{\theta_\pi}{\mathrm{argmax}} \, \mathbb{E}\left[\mathscr{L}^{\mathrm{clip}}(s, a; \theta_\pi)\right]. \tag{14.38}$$

The details of our PPO algorithm-based technique for joint trajectory design, transmission power, EH time, and phase shift matrix optimisation in RIS-assisted UAV communications are presented in Algorithm 14.2.

## 14.6 Simulation results

In our work, we use the Tensorflow 1.13.1 [42] for implementing our algorithms. We deploy the UAV at $(0, 0, 200)$, the RIS at $(200, 0, 50)$ and assume $d/\lambda = 1/2$ for convenience. All other parameters are provided in Table 14.3. In order to compare our proposed model with other baseline schemes, in this chapter, we consider the techniques as follows:

- **Optimisation with the hovering UAV**: the UAV is maintained at a fixed position at the centre of the cluster $(0, 0, H_{\mathrm{UAV}})$. We optimise the EH time $\tau$, the transmission power $P_0$, and the phase shift matrix at the RIS. We use the DDPG algorithm (H-DDPG) and the PPO algorithm (H-PPO) to solve the problem in the hovering UAV scenario.

*Table 14.3  Simulation parameters*

| Parameters | Value |
|---|---|
| Bandwidth ($W$) | 1 MHz |
| Maximum UAV transmission power | 5 W |
| UAV maximum speed per time step | 5 m |
| UAV's coverage | 500 m |
| The initial UAV's position | (0, 0, 200) |
| The RIS's position | (200, 0, 50) |
| Path-loss parameter | $\kappa_1 = 4, \kappa_2 = 2, \kappa_3 = 2.2$ |
| Channel power gain | $\beta_0 = -30$ dB |
| Rician factor | $\beta_1 = 4$ |
| Noise power | $\alpha^2 = -134$ dBm |
| Max. number of episodes | $M = 2000$ |
| Max. number of time step | $T = 400$ |
| Clipping parameter | $\varepsilon = 0.2$ |
| Discounting factor | $\zeta = 0.9$ |
| Max. number of IoT devices | 20 |
| Initial batch size | $K = 32$ |

- **Our proposed model with mobile UAV**: For the game formulated as in Section 14.4, we use the DDPG algorithm (F-DDPG) and the PPO algorithm (F-PPO) for solving the problem of joint optimisation of trajectory, power allocation, EH time scheduling and the phase shift matrix at the RIS.
- **Random selection scheme (RSS)**: The value of $\Phi$ is selected randomly and we use the DDPG algorithm (RSS-HDDPG) for optimising the EH time $\tau$ and the power $P_0$ in the hovering UAV scenario.
- **Without RIS**: We do not deploy the RIS in this scenario and optimise the EH time $\tau$, the power $P_0$ in the hovering UAV scenario using the PPO algorithm (WithoutRIS-HPPO).
- **Fixed EH time (FEH)**: The EH time $\tau$ is fixed to 0.5 and the flying path, power allocation $P_0$ and the phase shift $\Phi$ are optimised to maximise the performance. We use the DDPG algorithm (FEH-DDPG) for optimisation.
- **Maximal transmission power (MTP)**: The transmission power at the UAV is set to $P_0 = P_{\max}$. We use the PPO algorithm (MTP-PPO) to optimise the UAV's trajectory, EH time scheduling, and the RIS's phase shift matrix.

First, we consider the hovering UAV scenario and compare the performance versus the different number of IoT devices, $N$ with the number of RIS, $K = 20$ in Figure 14.2. We take the average of over 500 episodes for each scheme to draw the figures. When using the H-PPO algorithm, the total expected throughput is higher than in other schemes, including the ones using the H-DDPG algorithm, the RSS-HDDPG and the Without RIS-HPPO technique. The results suggest that with the EH time and the RIS's reflecting coefficient optimisation, the PPO algorithm is adequate irrespectively of the number of IoT devices.

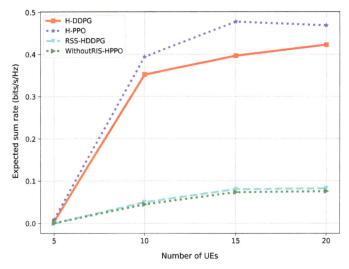

*Figure 14.2    Sum-rate performance in the hovering UAV scenario with different numbers of IoT devices, N. © IEEE 2022. Reprinted with permission from [1].*

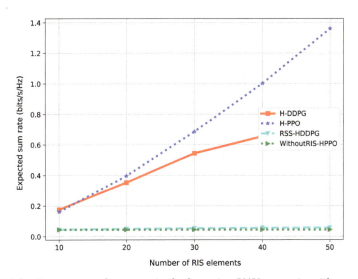

*Figure 14.3    Sum-rate performance in the hovering UAV scenario with varying number of RIS elements, K. © IEEE 2022. Reprinted with permission from [1].*

Next, we present the achieved sum rate of the PPO and DDPG algorithm in the hovering UAV scenario comparing with the RSS and without RIS case while the number of IoT devices is fixed at $N = 10$ in Figure 14.3. The H-PPO again shows the effective results with the different number of RIS elements, $K$. The sum-rate

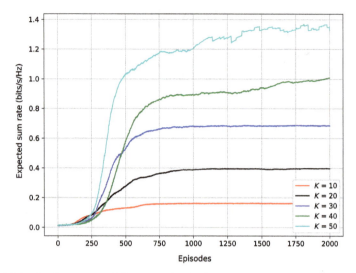

*Figure 14.4  Performance while using the PPO algorithm in the hovering UAV scenario with different number of RIS elements, K. © IEEE 2022. Reprinted with permission from [1].*

performance of the H-PPO algorithm improves from 0.2 to 1.4 (bits/s/Hz) following the increase of the RIS elements. The RIS is a passive reflector; thus, the reflected signal is diverse and not towards the destinations if we cannot control the coefficient of the RIS and select the phase shift randomly.

Furthermore, in Figure 14.4, we investigate the performance while using the PPO algorithm in the hovering UAV scenario with different numbers of RIS elements, $K$. When we increase $K$, the performance is enhanced. The results suggest that the PPO algorithm can handle large optimising variables and is effective in the hovering UAV scenario.

In Figure 14.5, we compare the performance of optimising for the mobile UAV to the hovering UAV with different number of IoT devices, $N$. When we optimise the UAV's trajectory, the performance is enhanced using both DDPG and PPO algorithms. The F-PPO approach reaches the best-expected result with different number of IoT devices, $N$.

In Figure 14.6, we present the advantage of optimising the UAV's trajectory in comparison to the hovering UAV scenario. While increasing the number of RIS elements, the performance is increased. The sum rate obtained by using the PPO algorithm is higher than the ones using the DDPG algorithm in both hovering UAV and mobile UAV scenarios. The results in Figures 14.5 and 14.3 suggest that the optimisation of the UAV's trajectory is effective.

In Figure 14.7, we compare the total sum rate in the mobile UAV with the number of RIS elements $K = 20$ and different numbers of IoT devices, $N$. The method based on the F-PPO algorithm shows impressive results over other schemes. When using the F-PPPO algorithm, we can achieve a total throughput of around 0.8 bits/Hz

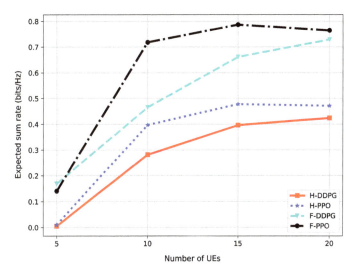

*Figure 14.5  The performance with different number of IoT devices, N. © IEEE 2022. Reprinted with permission from [1].*

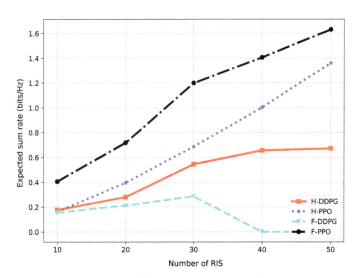

*Figure 14.6  Performance with different number of RIS elements, K. © IEEE 2022. Reprinted with permission from [1].*

with number of IoT devices set to 20. The performance obtained by the MTP-PPO algorithm is slightly lower than the F-PPO algorithm. Moreover, the FEH-DDPG and F-DDPG algorithms are not good and unstable. The reason is that the exploration–exploitation scheme for the DDPG algorithm is based on the initial value and the training scheme for the trajectory problem is affected by the noise of the training samples.

### 236 DRL for RIS and UAV Empowered Smart 6G Communications

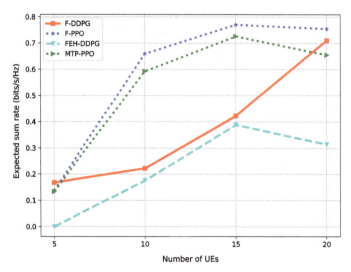

*Figure 14.7   Sum-rate performance with different number of IoT devices, N.
© IEEE 2022. Reprinted with permission from [1].*

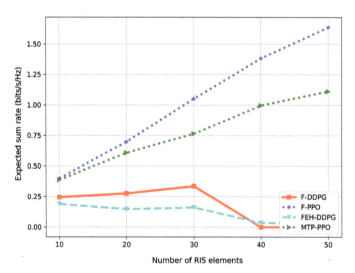

*Figure 14.8   Sum-rate performance with different numbers of RIS elements, K.
© IEEE 2022. Reprinted with permission from [1].*

We consider different numbers of RIS elements $K$ and compare the performance of our proposed algorithms with REH schemes and hovering UAV scenarios, as shown in Figure 14.8. The F-PPO algorithm-based technique outperforms other schemes while it reaches around 1.6 (bits/s/Hz). Following the performance using the F-PPO algorithm is the ones using the MTP-PPO algorithm. The achievable sum rate when we optimise the UAV's trajectory, transmission power, EH time and phase shift matrix is higher than when we only consider the fixed EH time in FEH or when we use the maximum power in MTP.

## 14.7 Summary

In this chapter, we have introduced a new system model for RIS-assisted UAV communications with the downlink power transfer and uplink information transmission protocol. We have proposed two DRL techniques for jointly optimising the UAV's trajectory, power allocation, IoT's EH time scheduling and the phase shift matrix of the RIS to maximise the network's throughput. The results suggest that the systems learned by the DRL algorithm can deal with large-scale scenarios and satisfy some power restrictions and processing time in RIS-assisted UAV communications. Future research directions include the possibility to consider a distributed model and cooperative communications with multiple UAVs.

## References

[1] Nguyen KK, Masaracchia A, Sharma V, *et al.* RIS-Assisted UAV Communications for IoT with Wireless Power Transfer using Deep Reinforcement Learning. *IEEE J Sel Topics Signal Process*. 2022;16(5):1086–1096.

[2] Nguyen LD, Kortun A, and Duong TQ. An Introduction of Real-time Embedded Optimisation Programming for UAV Systems under Disaster Communication. *EAI Endorsed Transactions on Industrial Networks and Intelligent Systems*. 2018;5(17):1–8.

[3] Nguyen KK, Vien NA, Nguyen LD, *et al.* Real-Time Energy Harvesting Aided Scheduling in UAV-Assisted D2D Networks Relying on Deep Reinforcement Learning. *IEEE Access*. 2021;9:3638–3648.

[4] Nguyen KK, Duong TQ, Do-Duy T, *et al.* 3D UAV Trajectory and Data Collection Optimisation via Deep Reinforcement Learning; 2021. Available from: https://arxiv.org/abs/2106.03129.

[5] Xie L, Xu J, and Zhang R. Throughput Maximization for UAV-Enabled Wireless Powered Communication Networks. *IEEE Internet Things J*. 2019;6(2):1690–1703.

[6] Chen Z, Chi K, Zheng K, *et al.* Minimization of Transmission Completion Time in UAV-Enabled Wireless Powered Communication Networks. *IEEE Internet Things J*. 2020;7(2):1245–1259.

[7] Xie L, Xu J, and Zeng Y. Common Throughput Maximization for UAV-Enabled Interference Channel With Wireless Powered Communications. *IEEE Trans Commun*. 2020;68(5):3197–3212.

[8] Hashir SM, Mehrabi A, Mili MR, *et al.* Performance Trade-Off in UAV-Aided Wireless-Powered Communication Networks via Multi-Objective Optimization. *IEEE Trans Veh Technol*. 2021;70(12):13430–13435.

[9] Nguyen MN, Nguyen LD, Duong TQ, *et al.* Real-Time Optimal Resource Allocation for Embedded UAV Communication Systems. *IEEE Wireless Commun Lett*. 2019;8(1):225–228.

[10] Nguyen KK, Duong TQ, Vien NA, et al. Distributed Deep Deterministic Policy Gradient for Power Allocation Control in D2D-Based V2V Communications. *IEEE Access*. 2019;7:164533–164543.

[11] Nguyen KK, Duong TQ, Vien NA, et al. Non-Cooperative Energy Efficient Power Allocation Game in D2D Communication: A Multi-Agent Deep Reinforcement Learning Approach. *IEEE Access*. 2019;7:100480–100490.

[12] Nguyen LD, Nguyen KK, Kortun A, et al. Real-Time Deployment and Resource Allocation for Distributed UAV Systems in Disaster Relief. In: *Proc. IEEE 20th International Workshop on Signal Processing Advances in Wireless Commun. (SPAWC)*. Cannes, France; 2019. p. 1–5.

[13] Basar E, Renzo MD, Rosny JD, et al. Wireless Communications Through Reconfigurable Intelligent Surfaces. *IEEE Access*. 2019;7:116753–116773.

[14] Atapattu S, Fan R, Dharmawansa P, et al. Reconfigurable Intelligent Surface Assisted Two-Way Communications: Performance Analysis and Optimization. *IEEE Trans Commun*. 2020;68(10):6552–6567.

[15] Yu H, Tuan HD, Nasir AA, et al. Joint Design of Reconfigurable Intelligent Surfaces and Transmit Beamforming Under Proper and Improper Gaussian Signaling. *IEEE J Select Areas Commun*. 2020;38(11):2589–2603.

[16] Li Y, Jiang M, Zhang Q, et al. Joint Beamforming Design in Multi-Cluster MISO NOMA Reconfigurable Intelligent Surface-Aided Downlink Communication Networks. *IEEE Trans Commun*. 2021;69(1):664–674.

[17] Ge L, Dong P, Zhang H, et al. Joint Beamforming and Trajectory Optimization for Intelligent Reflecting Surfaces-Assisted UAV Communications. *IEEE Access*. 2020;8:78702–78712.

[18] Li S, Duo B, Yuan X, et al. Reconfigurable Intelligent Surface Assisted UAV Communication: Joint Trajectory Design and Passive Beamforming. *IEEE Wireless Commun Lett*. 2020;9(5):716–720.

[19] Ranjha A and Kaddoum G. URLLC Facilitated by Mobile UAV Relay and RIS: A Joint Design of Passive Beamforming, Blocklength, and UAV Positioning. *IEEE Internet Things J*. 2021;8(6):4618–4627.

[20] Nguyen KK, Khosravirad S, Costa DBD, et al. Reconfigurable Intelligent Surface-assisted Multi-UAV Networks: Efficient Resource Allocation with Deep Reinforcement Learning. *IEEE J Selected Topics in Signal Process*. 2021;1–1. Early Access.

[21] Zou Y, Gong S, Xu J, et al. Wireless Powered Intelligent Reflecting Surfaces for Enhancing Wireless Communications. *IEEE Trans Veh Technol*. 2020;69(10):12369–12373.

[22] Huang C, Mo R, and Yuen C. Reconfigurable Intelligent Surface Assisted Multiuser MISO Systems Exploiting Deep Reinforcement Learning. *IEEE J Select Areas Commun*. 2020;38(8):1839–1850.

[23] Pan C, Ren H, Wang K, et al. "Intelligent reflecting surface aided MIMO broadcasting for simultaneous wireless information and power transfer," *IEEE J. Select. Areas Commun.*, vol. 38, no. 8, pp. 1719–1734, Aug. 2020.

[24] Yang H, Yuan X, Fang J, et al. Reconfigurable Intelligent Surface Aided Constant-Envelope Wireless Power Transfer. *IEEE Trans Signal Process*. 2021;69:1347–1361.

[25] Lin S, Zheng B, Alexandropoulos GC, *et al*. Reconfigurable Intelligent Surfaces with Reflection Pattern Modulation: Beamforming Design and Performance Analysis. *IEEE Trans Wireless Commun*. 2021;20(2): 741–754.
[26] Chen Y, Ai B, Zhang H, *et al*. Reconfigurable Intelligent Surface Assisted Device-to-Device Communications. *IEEE Trans Wireless Commun*. 2021;20(5):2792–2804.
[27] Wang L, Wang K, Pan C, *et al*. *Joint Trajectory and Passive Beamforming Design for Intelligent Reflecting Surface-Aided UAV Communications: A Deep Reinforcement Learning Approach*; 2020. Available from: https://arxiv.org/abs/2007.08380.
[28] Yang H, Xiong Z, Zhao J, *et al*. Deep Reinforcement Learning-Based Intelligent Reflecting Surface for Secure Wireless Communications. *IEEE Trans Wireless Commun*. 2021;20(1):375–388.
[29] Feng K, Wang Q, Li X, *et al*. Deep Reinforcement Learning Based Intelligent Reflecting Surface Optimization for MISO Communication Systems. *IEEE Wireless Commun Lett*. 2020;9(5):745–749.
[30] Nguyen KK, Khosravirad S, Costa DBD, *et al*. Reconfigurable Intelligent Surface-assisted Multi-UAV Networks: Efficient Resource Allocation with Deep Reinforcement Learning. *IEEE J Select Topics Signal Process*. 2021; 1–1. Early access.
[31] Yin S, Zhao S, Zhao Y, *et al*. Intelligent Trajectory Design in UAV-Aided Communications with Reinforcement Learning. *IEEE Trans Veh Technol*. 2019;68(8):8227–8231.
[32] Yang Z, Chen M, Saad W, *et al*. Energy-Efficient Wireless Communications with Distributed Reconfigurable Intelligent Surfaces. *IEEE Trans Wireless Commun*. 2021; Early Access.
[33] Liu X, Liu Y, and Chen Y. Machine Learning Empowered Trajectory and Passive Beamforming Design in UAV-RIS Wireless Networks. *IEEE J Select Areas Commun*. 2021;39(7):2042–2055.
[34] Cao X, Yang B, Huang C, *et al*. Reconfigurable Intelligent Surface-Assisted Aerial-Terrestrial Communications via Multi-Task Learning. *IEEE J Select Areas Commun*. 2021;39(10):3035–3050.
[35] Bertsekas DP. Dynamic Programming and Optimal Control. vol. 1. *Athena Scientific Belmont*, MA; 1995.
[36] X. Yu, D. Xu, Y. Sun, D. W. K. Ng, and R. Schober, "Robust and secure wireless communications via intelligent reflecting surfaces," *IEEE J. Select. Areas Commun.*, vol. 38, no. 11, pp. 2637–2652, 2020.
[37] Lillicrap T. P, Hunt J. J, Pritzel A, *et al*. "Continuous control with deep reinforcement learning," in *Proc. 4th International Conf. on Learning Representations (ICLR)*, 2016.
[38] Kingma DP and Ba JL. *Adam: A Method for Stochastic Optimization*; 2014. Available from: https://arxiv.org/abs/1412.6980.
[39] Schulman J, Moritz P, Levine S, *et al*. High-Dimensional Continuous Control Using Generalized Advantage Estimation. In: *Proc. 4th International Conf. Learning Representations (ICLR)*; 2016.

[40] Schulman J, Wolski F, Dhariwal P, *et al*. *Proximal Policy Optimization Algorithms*; 2017.
[41] Mnih V, Badia A. P, Mirza M, *et al*. "Asynchronous methods for deep reinforcement learning," in *Proc. Int. Conf. Mach. Learn. PMLR*, 2016, pp. 1928–1937.
[42] Abadi M, Barham P, Chen J, *et al*. "Tensorflow: A system for large-scale machine learning," in *Proc. 12th USENIX Sym. Opr. Syst. Design and Imp. (OSDI 16)*, Nov. 2016, pp. 265–283.

*Chapter 15*
# Multi-agent learning in networks supported by RIS and multi-UAVs

In this chapter, we illustrate a method for maximising the energy efficiency (EE) of a RIS-assisted UAVs-based network by jointly optimising the UAVs' power allocation and the RIS's phase-shift matrix through a DRL approach. Moreover, the parallel learning approach is also proposed as an effective possible solution for reducing the information transmission requirement of the centralised approach. Numerical results show a significant improvement in our proposed schemes compared with the conventional approaches in terms of EE, flexibility, and full suitability for real-time applications. *

## 15.1 Introduction

Unmanned aerial vehicles (UAVs) have recently been widely applied in numerous fields due to their agility. The high altitude of UAVs can overcome some bottlenecks of the existing scenarios, such as building blockage, remote areas, and emergency services. Some real-life applications of the UAVs are surveillance [2], geography exploration [3], disaster rescue mission [4–6], and wireless communications [7,8]. The UAVs are also playing a crucial role in bringing beyond the fifth generation (5G) network to every corner of the world owing to their low-cost production and flexibility. Ultimately, UAV-assisted wireless networks significantly enhance the network's coverage and improve information transmission efficiency.

Very recently, reconfigurable intelligent surface (RIS) has emerged as a cutting-edge technology for beyond 5G and sixth generation (6G) networks. In particular, a massive number of reflective elements are intelligently controlled to reflect the received signal towards the destinations. The controller helps the RIS be dynamically adapted to the propagation environment with the aim to meet different purposes; for example, enhance the arrival signal and mitigate the interference [9–16]. The RIS has been recently deployed efficiently due to its low-cost hardware production, nearly passive nature, easy deployment, communication without new waves, and energy-saving nature.

Owing to the intrinsic features of RIS and UAVs, RIS-assisted UAV communication has been recently considered for enhancing network performance. Although

*This chapter has been published partly in [1].

the high altitude of the UAV significantly strengthens the channel between the UAV and the users, the connections are sometimes blocked by buildings or other obstacles in specific scenarios. Thus, the RIS attached to the building or on a high place can reflect the channel from the UAV to the users [17–19]. Moreover, the data through the RIS will experience fewer intermediate delays and more freshness than when we use a mobile active relay in the middle. On the other hand, the RIS is easy to deploy and effective in reducing power consumption.

Deep reinforcement learning (DRL) algorithms have emerged as a powerful method for embedded optimisation and instant decision-making models in wireless networks. The DRL methods have been used for device-to-device (D2D) communication [20,21], UAV-assisted networks [8], and RIS-assisted wireless networks [22]. Inspired by the impressive results, in this chapter, we use the DRL algorithm to enable the UAVs to choose the proper power level and to adjust the phase-shift matrix of the RIS to maximise the reward. The neural networks are trained in the offline phase and then deployed in the terminal devices or controllers. The proper actions can be chosen in milliseconds or instants in a centralised and decentralised manner to obtain optimal performance in the multi-UAVs-assisted networks with the support of RIS.

### 15.1.1 Related work

The high-flying altitude of the UAV helps the wireless networks improve the coverage and transmit signal [5,7,8,23]. In [5], multiple UAVs were deployed in a disaster area to efficiently support the users. The K-means algorithm was proposed for the deployment mission, while the block coordinate descent (BCD) procedure was used to maximise the worst end-to-end sum rate. In [7], the authors used the UAV as a mobile data collector. The optimal UAV's flying path and the wake-up scheduling at the sensor nodes helped to reduce the energy consumption in both the UAV and the sensors. The authors in [8] considered the UAV as an energy supplier for the non-fixed power source devices to assist communications in D2D networks. In [23], the UAV's trajectory was optimised to maximise the energy efficiency (EE) in an unconstrained condition and circular trajectory.

As aforementioned, RIS has been recently attracting enormous attention as an emerging technology for enabling beyond 5G due to its unique characteristics, which include low-cost production and less energy consumption [9–16,24,25]. In [9], an algorithm was proposed for maximising the weighted sum – of all users via beamforming vector and RIS phase-shift optimisation under the perfect channel state information (CSI) and imperfect CSI scenarios. In [12], the power allocation and the phase-shift optimisation algorithm were proposed for maximising the EE performance. In [11], the RIS was used for enhancing communication and reducing interference in the D2D networks. Two sub-problems with the fixed power transmission and the discrete RIS's phase-shift matrix were considered and solved efficiently. The authors in [14] optimised the beamforming vector at the secondary user transmitter and the RIS phase shift in a downlink multiple-input single-output (MISO) cognitive radio system with multiple RISs. The perfect CSI and imperfect CSI scenarios were considered; then, the BCD procedure was used to maximise the achievable sum rate.

By utilising both the advantages of the UAV and the RIS, the network performance is significantly improved through enhancing the received signal and mitigating the interference [17–19]. In [17], the joint beamforming vector, trajectory and phase-shift optimisation algorithm were proposed for maximising the received signal at the ground users in the UAV-assisted wireless communications. In [18], the joint UAV flying path and RIS passive beamforming design were investigated in order to maximise the network sum rate. Two sub-problems with the fixed trajectory and the optimal phase-shift matrix were solved using a closed-form solution and the successive convex approximation method. The UAV communications supported by RIS have been extended to the concept of ultra-reliable and low-latency communication (URLLC) in [19] where the RIS passive beamforming, the UAV's position and URLLC blocklength were optimised for minimising the total decoding error rate of URLLC.

Since DRL is an effective solution for solving the dynamic environment with continuous moving [8,20,21,26], some recent works have explored the efficiency of the DRL techniques for RIS-assisted wireless networks [22,27–29]. In [22], a DRL algorithm was proposed for optimising the RIS phase shift in order to maximise the signal-to-noise ratio (SNR). The authors in [28] optimised the transmit beamforming vector and the RIS phase-shift model by using the DRL algorithm to maximise the total sum rate. A deep Q-learning and deep deterministic policy gradient were proposed and showed impressive results in the MISO communications. To minimise the sum of age-of-information, the authors in [29] proposed a DRL algorithm to adjust the UAV's altitude and the RIS phase shift. However, these techniques mostly assume idealistic conditions or flat fading channel settings.

## 15.1.2 Contributions

However, when deploying the optimisation algorithm with DRL into RIS-assisted UAV communications, previous works assumed the perfect condition of the environment, flat fading channels, static users, and perfect CSI, which are unrealistic and infeasible for real-life applications. Furthermore, the delay when using a mathematical model in centralised learning is huge for real-time use cases. To overcome these aforementioned shortcomings, in this chapter, we propose efficient DRL algorithms by jointly optimising the power allocation of the UAV and the RIS's phase shifts for maximising the EE and the network's sum rate. The DRL approaches bring a flexible and autonomous ability to UAVs and RIS. With trained neural networks, the UAVs and RIS can choose a proper action without delay. Furthermore, continuous learning with up-to-date data by interaction with the environment helps the UAVs and RIS to adapt to the dynamic environment. To the best of our knowledge, our work is the first technical chapter that exploits the efficiency of DRL techniques in multi-UAV-assisted wireless communications with the support of RISs. The main contributions of this chapter can be summarised as follows:

- We conceive a wireless network of multi-UAVs supported by an RIS. Each UAV is deployed to serve a specific cluster of UEs. Due to the severe shadowing effect, the RIS is used to enhance the received signal's quality at the UEs from the associated UAV and to mitigate interference from others.

- The EE problem is formulated for the downlink channel with the power restrictions and the RIS's requirement. To optimise the EE network performance, we propose a centralised DRL technique for jointly solving the power allocation at the UAVs and phase-shift matrix of the RIS. Then, parallel learning is used to train each element in our model to be intelligent and to reduce the delay when transmitting the action between the UAV and the RIS.
- To improve the network performance, we introduce the proximal policy optimisation (PPO) algorithm with a better sampling technique.
- Through the numerical results, we demonstrate that our proposed methods efficiently solve the joint optimisation problem within dynamic environments and time-varying CSI and outperform the other benchmarks.

The remainder of this chapter is organised as follows. We present the system model and problem formulation for the energy-efficient multi-UAV-assisted wireless communications with the support of the RIS in Section 15.2. The mathematical backgrounds for the DRL algorithm are presented in Section 15.3. The centralised DDPG approach for joint optimisation of power allocation and phase shift in multi-UAV-assisted wireless networks is introduced in Section 15.4. We propose parallel learning for our approach to reduce delay in Section 15.5. Moreover, the PPO algorithm is proposed for solving both centralised and decentralised learning in Section 15.6. Numerical results are illustrated in Section 15.7 while the conclusion and future works are presented in Section 15.8.

## 15.2 System model and problem formulation

As illustrated in Figure 14.1, we consider a downlink multi-UAV wireless network assisted by one RIS. Each UAV is equipped with a single antenna for serving a specific cluster of a group of users (UEs), in which it is assumed $N$ UAVs correspond to $N$ clusters of UEs, where each cluster consists of $M$ single-antenna UEs. The UEs are randomly distributed in the coverage $C$ from the centre of each cluster. The channel between the UAV and UEs is blocked by the building, wall and concretes. Thus, we deploy a RIS with $K$ elements to support the information transmission from UAVs to UEs.

### 15.2.1 System model

We assume that the coordinate of the $n$th UAV and $m$th UEs in the $n$th cluster at the time step $t$ is $X_n^t = (x_n^t, y_n^t, H_n^t)$ and $X_{mn}^t = (x_{mn}^t, y_{mn}^t)$, with $n = 1, \ldots, N$ and $m = 1, \ldots, M$ (see Figure 15.1). The RIS is attached at the building or a high location at $(x^t, y^t, z^t)$, respectively.

The distance between the $n$th UAV and the RIS panel in time step $t$ is denoted by:

$$d_n^t = \sqrt{(x_n^t - x^t)^2 + (y_n^t - y^t)^2 + (H_n^t - z^t)^2}. \tag{15.1}$$

Similarly, the distance between the RIS panel and the $m$th UEs in the $n$th cluster is written as:

$$d_{nm}^t = \sqrt{(x^t - x_{nm}^t)^2 + (y^t - y_{nm}^t)^2 + (z^t)^2}. \tag{15.2}$$

## Multi-agent learning in networks supported by RIS and multi-UAVs    245

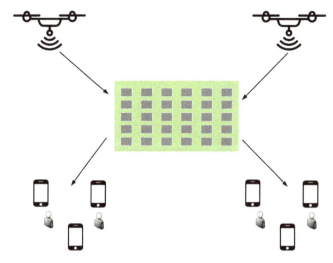

*Figure 15.1   System setup. ©IEEE 2022. Reprinted with permission from [1]*

Due to the high shadowing and severe blocking effect, the direct links between UAVs and UEs do not exist, and therefore, it is only considered the alternative paths (reflected links) via RIS's reflection. The links between the UAVs and the RIS are modelled as air-to-air (AA) channels, whereas the link between the RIS and the UEs is assumed to follow an air-to-ground (AG) channel. Following the AA channel model, the channel gain between the $n$th UAV and the RIS in time step $t$ is formulated as:

$$H_{n,RIS}^t = \sqrt{\beta_0 (d_n^t)^{-\kappa_1}} \left[1, e^{-j\frac{2\pi}{\lambda} d \cos(\phi_{AoA}^t)}, \ldots, e^{-j\frac{2\pi}{\lambda}(K-1)d \cos(\phi_{AoA}^t)}\right]^T, \quad (15.3)$$

where $\kappa_1$ is the path loss exponent for the UAV-RIS link, $d$ is element spacing, and $\lambda$ is the carrier wavelength; the right term of (15.3) is the signal from the $n$th UAV to the RIS, $\cos(\phi_{AoA}^t)$ is the cosine of the angle of arrival (AoA).

According to the AG channel model, the channel gain between the RIS and the $m$th UEs in the $n$th cluster can be written as:

$$h_{RIS,nm}^t = \sqrt{\beta_0 (d_{nm}^t)^{-\kappa_2}} \left(\sqrt{\frac{\beta_1}{1+\beta_1}} h_{RIS,nm}^{LoS} + \sqrt{\frac{1}{\beta+1}} h_{RIS,nm}^{NLoS}\right), \quad (15.4)$$

where the deterministic LoS component is denoted by $h_{RIS,nm}^{LoS} = [1, e^{-j\frac{2\pi}{\lambda} d \cos(\phi_{AoD}^t)}, \ldots, e^{-j\frac{2\pi}{\lambda}(K-1)d \cos(\phi_{AoD}^t)}]$ and the non-light-of-sight (NLoS) component is modelled as complex Gaussian distribution with a zero-mean and unit-variance $\mathcal{CN}(0, 1)$; $\cos(\phi_{AoD})$ is the angle of departure (AoD) from the RIS to the $m$th UE in the $n$th cluster; $\beta_1$ and $\kappa_2$ are the Rician factor and the path loss exponent for the RIS-UEs link, respectively.

The signal from the UAV to UEs is reflected by the RIS. Thus, the received signal from the $n$th UAV to the $m$th UE in the $n$th cluster at time step $t$ can be written as:

$$y_{nm}^t = H_{n,RIS}^t \Phi^t h_{RIS,nm}^t \sqrt{P_n} x + \eta, \quad (15.5)$$

where $H_n^t \in \mathbb{C}^{1 \times K}$ is the channel gains array from the $n$th UAV to the RIS, $\eta$ is the power noise signal following the complex Gaussian distribution with power $\alpha^2$; $P_n$ and $x$ are the transmit power and the symbol signal sent from the $n$th UAV, respectively; $\Phi^t = \text{diag}[\phi_1^t, \phi_2^t, \ldots, \phi_K^t]$ is the diagonal matrix at the RIS, where $\phi_k^t = e^{j\theta_k^t}, \forall k = 1, 2, \ldots, K$ with $\theta_k^t \in [0, 2\pi]$ is the phase shift of the $k$th element in the RIS at time step $t$.

### 15.2.2 Problem formulation

In this work, we consider downlink communications where the signal from the UAV is dedicated to a designated UE in the associated cluster. In other words, the $m$th UE in the $n$th cluster receives the information from the $n$th UAV while the signals from other UAVs are considered as interference. Thus, the received signal-to-interference-plus-noise-ratio (SINR) at the $m$th UE in the cluster $n$ at time step $t$ can be formulated as follows:

$$\gamma_{nm}^t = \frac{P_n^t |H_{n,RIS}^t \Phi^t h_{RIS,nm}^t|^2}{\sum_{i \neq n}^N P_i^t |H_{i,RIS}^t \Phi^t h_{RIS,im}^t|^2 + \alpha^2}, \tag{15.6}$$

The throughput at the $m$th UEs in the $n$th cluster at time step $t$ is:

$$R_{nm}^t = B \log_2 (1 + \gamma_{nm}^t), \tag{15.7}$$

where $B$ is the bandwidth. The total throughput at time step $t$ is cumulative from the UEs of all clusters, and it can be expressed by:

$$R_{total}^t = \sum_{n=1}^N \sum_{m=1}^M R_{nm}^t, \tag{15.8}$$

and the total power consumption is given by:

$$P_{total} = \sum_{n=1}^N P_n + P_K + P_c, \tag{15.9}$$

where $P_K$ and $P_c$ are the power consumption at the RIS and the power circuit at the UAV, respectively.

Our objective is to maximise the EE of all UEs by jointly optimising the transmit powers at the UAVs and the phase shifts at the RIS. In each time step $t$, each UAV will choose the proper power, and each RIS element will choose the phase-shift value depending on the local information that each component receives from the environment. The optimisation of maximising the EE of all UEs subject to the transmit power at UAVs and phase shifts of RIS can be formulated as:

$$\max_{P, \Phi} \quad \frac{\sum_{n=1}^N \sum_{m=1}^M R_{nm}^t}{\sum_{n=1}^N P_n + P_K + P_c}$$
$$\text{s.t.} \quad 0 \leq P_n \leq P_{max}, \forall n \in N, \tag{15.10}$$
$$\theta_k \in [0, 2\pi], \forall k \in K,$$

where $P = \{P_1, \ldots, P_N\}$ and $P_{max}$ are the vector of power and the maximum information transmission power at the UAVs, respectively. To solve the maximised EE problem, we propose two DRL algorithms for the centralised approach, and then the parallel learning distributed approach is introduced for practical applications.

## 15.3 Preliminaries

To deploy a system with the support of the DRL algorithms, we have two main approaches: value search and policy search. In the value search approach, we consider the gap between the received reward in two samples to adjust the value function. In the policy search algorithm, we directly find the policy for the problems. We represent the Markov Decision Process (MDP) [30] by $< \mathscr{S}, \mathscr{A}, \mathscr{P}, \mathscr{R}, \zeta >$, where $\mathscr{S}, \mathscr{A}$ denote the agent's state space and action space; $\mathscr{P}_{ss'}(a)$ denotes the state transition probability with $s = s^t, s' = s^{t+1} \in \mathscr{S}, a \in \mathscr{A}$; $r \in \mathscr{R}$ is the reward function; and $\zeta$ is the discount factor.

### 15.3.1 Value function

The idea of the value function methods relies on the estimation of the value in a given state. The state-value function $V^\pi(s)$ is obtained following the policy $\pi$ starting at the state $s$ as:

$$V^\pi = \mathbb{E}\{\mathscr{R}|s, \pi\}, \tag{15.11}$$

where $\mathbb{E}\{\cdot\}$ is the expectation operation that depends on the transition function $\mathscr{P}_{ss'(a)} = p(s'|s, a)$ and the stochastic property of the policy $\pi$.

Our goal is to find the optimal policy $\pi^*$, which has a corresponding optimal state-value function $V^*(s)$ as:

$$V^*(s) = \max_\pi V^\pi(s), s \in \mathscr{S}. \tag{15.12}$$

To maximise the expected cumulative reward, the agent chooses the action $a \in \mathscr{A}$ following the optimal policy $\pi^*$ that satisfies the Bellman align [30]:

$$V^*(s) = V^{\pi^*} = \max_{a \in \mathscr{A}} \left\{ \mathbb{E}(r(s, a)) + \zeta \sum_{s' \in \mathscr{S}} P_{ss'}(a) V^*(s') \right\}. \tag{15.13}$$

The action-value function is defined as the obtained reward when the agent takes action $a$ at the state $s$ under the policy $\pi$ as:

$$Q^\pi(s, a) = \mathbb{E}(r(s, a)) + \zeta \sum_{s' \in \mathscr{S}} P_{ss'}(a) V(s'). \tag{15.14}$$

The optimal policy $Q^*(s, a) = Q^{\pi^*}$, we have:

$$V^*(s) = \max_{a \in \mathscr{A}} Q^*(s, a). \tag{15.15}$$

## 15.3.2 Policy search

Instead of considering the value function model, the agent can directly find an optimal policy $\pi^*$. Among policy search methods, the policy gradient is most popular due to its efficient sampling with a large number of parameters. The reward function is defined by the performance under the policy $\pi$ as:

$$J(\theta) = \sum_{s\in\mathscr{S}} d^\pi(s) \sum_{a\in\mathscr{A}} \pi_\theta(a|s) r^\pi(s,a), \tag{15.16}$$

where $\theta_\pi$ is the vector of the policy parameters and $d^\pi(s)$ is the stationary distribution of Markov chain with the policy $\pi_\theta$. The optimal policy $\pi^*$ can be obtained by using gradient ascent for adjusting the parameters $\theta_\pi$ relying on the $\nabla_\theta J(\theta_\pi)$. For any MDP, we have [31]:

$$\begin{aligned}\nabla_\theta J &= \sum_{s\in\mathscr{S}} d^\pi(s) \sum_{a\in\mathscr{A}} \nabla_\theta \pi(a|s) Q^\pi(s,a) \\ &= \mathbb{E}_{\pi_\theta}\left[\nabla_\theta \ln \pi_\theta(s,a) Q^\pi(s,a)\right].\end{aligned} \tag{15.17}$$

The REINFORCE algorithm, a Monte-Carlo policy gradient learning, adjusts the parameters $\theta_\pi$ by estimating the return using Monte-Carlo methods and episode samples. The optimal policy parameter $\theta_\pi^*$ can be obtained by:

$$\theta_\pi^* = \underset{\theta_\pi}{\mathrm{argmax}}\,\mathbb{E}\left[\sum_a \pi(a|s;\theta_\pi) r(s,a)\right], \tag{15.18}$$

The gradient is defined as:

$$\nabla \theta_\pi = \mathbb{E}_\pi\left[\nabla_{\theta_\pi} \ln \pi(a|s;\theta_\pi) r(s,a)|_{s=s^t,a=a^t}\right]. \tag{15.19}$$

We use the gradient ascent to update the parameters $\theta_\pi$ as:

$$\theta_\pi \leftarrow \theta_\pi + \varepsilon \nabla \theta_\pi, \tag{15.20}$$

where $0 \leq \varepsilon \leq 1$ is the step-size parameter. The optimal action $a^*$ can be obtained with the maximum probability as follows:

$$a^* = \underset{a}{\mathrm{argmax}}\,\pi(a|s;\theta_\pi). \tag{15.21}$$

## 15.4 Centralised optimisation for power allocation and phase-shift matrix

In the centralised approach, we assume that the information is processed at a central point (e.g. cloud server), and the next action for each element in the system will be transferred at the beginning of each time step. Thus, to jointly optimise the power allocation at the UAVs and the phase-shift matrix at the RIS, we consider the central processing point to be an agent. The optimisation problem can be formulated by the MDP $<\mathscr{S},\mathscr{A},\mathscr{P},\mathscr{R},\zeta>$. Particularly, with our centralised optimisation, we formulate the game as follows:

- *State space*: The agent interacts with the environment for maximising the EE performance. Thus, the agent only has knowledge about the local information, e.g. the reflected channel gains. The state space is defined as follows:

$$\mathscr{S} = \{H_{1,RIS} \Phi h_{RIS,11}, H_{1,RIS} \Phi h_{RIS,12}, \\ \ldots, H_{n,RIS} \Phi h_{RIS,nm}, \ldots, H_{N,RIS} \Phi h_{RIS,NM}\}. \tag{15.22}$$

- *Action space*: With the downlink transmission in the RIS-assisted multi-UAV networks, we optimise the power allocation at UAVs and phase-shift matrix at RIS. Thus, the action space is defined as follows:

$$\mathscr{A} = \{P_1, P_2, \ldots, P_N, \theta_1, \theta_2, \ldots, \theta_K\}. \tag{15.23}$$

The agent takes the action $a^t = \{P_1^t, P_2^t, \ldots, P_N^t, \theta_1^t, \theta_2^t, \ldots, \theta_K^t\}$ at the state $s^t$ and moves to the next state $s' = s^{t+1}$.

- *Reward function*: Our objective is to maximise the EE performance; thus, we formulate the reward function as:

$$\mathscr{R} = \frac{\sum_{n=1}^{N} \sum_{m=1}^{M} R_{nm}^t}{\sum_{n=1}^{N} P_n + P_K + P_c}. \tag{15.24}$$

After formulating the EE game, we proposed a DRL algorithm for the agent to interact with the environment to find the optimal policy $\pi^*$. Deep deterministic policy gradient (DDPG) is a hybrid model composed of the actor part based on the value function and the critic component based on the policy search. In the DDPG algorithm, we use *experience replay buffer* and *target network* techniques to improve the convergence speed and avoid excessive calculation. In the *experience replay buffer*, we use a finite size of a memory size $B$ to store the executed transition $< s^t, a^t, r^t, s^{t+1} >$. After collecting enough samples, we randomly select a mini-batch $D$ of transitions from buffer $B$ for training the neural networks. The memory $B$ is set to a finite size for updating the new sample and discarding the old ones. Otherwise, we use *target networks* for the critic and actor network when calculating the target value.

We denote the critic network as $Q(s, a; \theta_q)$ with the parameter $\theta_q$ and the target critic network as $Q'(s, a; \theta_{q'})$ with the parameter $\theta_{q'}$. Similarly, we initialise the actor network $\mu(s; \theta_\mu)$ with the parameter $\theta_\mu$ and the target actor network $\mu'(s; \theta_{\mu'})$ with the parameter $\theta_{\mu'}$. We train the actor and critic network using the stochastic gradient descent (SGD) over a mini-batch of $D$ samples. The critic network is updated by minimising the quantity:

$$L = \frac{1}{D} \sum_i^D \left(y^i - Q(s^i, a^i; \theta_q)\right)^2, \tag{15.25}$$

with the target

$$y^i = r^i(s^i, a^i) + \zeta Q'(s^{i+1}, a^{i+1}; \theta_{q'})|_{a^{i+1} = \mu'(s^{i+1}; \theta_{\mu'})}. \tag{15.26}$$

The actor-network parameters are updated by:

$$\nabla_{\theta_\mu} J \approx \frac{1}{D} \sum_i^D \nabla_{a^i} Q(s^i, a^i; \theta_q)|_{a^i = \mu(s^i)} \nabla_{\theta_\mu} \mu(s^i; \theta_\mu). \tag{15.27}$$

The target actor network parameters $\theta_q$ and the target critic network parameters $\theta_{\mu'}$ are updated by using soft target updates as follows:

$$\theta_{q'} \leftarrow \varkappa\theta_q + (1 - \varkappa)\theta_{q'}, \tag{15.28}$$

$$\theta_{\mu'} \leftarrow \varkappa\theta_\mu + (1 - \varkappa)\theta_{\mu'}. \tag{15.29}$$

where $\varkappa$ is a hyperparameter between 0 and 1.

In the DDPG algorithm, the deterministic policy is trained in an off-policy way; thus, for *explorations* and *exploitations* purpose, we add a noise process of $\mathcal{N}(0, 1)$ as follows [32]:

$$\mu'(s^t; \theta^t_{\mu'}) = \mu(s^t; \theta^t_\mu) + \psi\mathcal{N}(0, 1) \tag{15.30}$$

where $\psi$ is a hyperparameter. The details of our DDPG algorithm-based technique for joint power allocation and phase-shift matrix optimisation in RIS-assisted UAV communications are presented in Algorithm 15.1, where $E$ and $T$ denote the number of the maximum episode and time step, respectively.

In Algorithm 15.1, the agent interacts with the environments to maximise the obtained reward (15.24). In time step $t$, the agent has local information of channel model $s^t \in \mathcal{S}$. At the state $s^t$, the agent chooses the action $a^t \in \mathcal{A}$ by the actor networks. By executing the action $a^t$ in the environment, the agent obtains a response following the reward function (15.24). Then, the critic and actor-network parameters are updated by training a mini-batch of $D$ transitions by stochastic gradient descent with Adam optimiser [33].

---

**Algorithm 15.1** Centralised optimisation for joint power allocation and phase-shift matrix in RIS-assisted UAV communications

1: Initialise the critic network $Q(s, a; \theta_q)$ and the target critic networks $Q'$
2: Initialise the actor-network $\mu(s; \theta_\mu)$ and the target actor network $\mu'$
3: Initialise replay memory pool $\mathcal{B}$
4: **for** episode = $1, \ldots, E$ **do**
5:     Initialise an action exploration process $\mathcal{N}$
6:     Receive initial observation state $s^0$
7:     **for** iteration = $1, \ldots, T$ **do**
8:         Execute the action $a^t$ obtained at state $s^t$
9:         Update the reward $r^t$ according to (15.24)
10:         Observe the new state $s^{t+1}$
11:         Store transition $(s^t, a^t, r^t, s^{t+1})$ into replay buffer $\mathcal{B}$
12:         Sample randomly a mini-batch of $D$ transitions $(s^i, a^i, r^i, s^{i+1})$ from $\mathcal{B}$
13:         Update critic parameter by stochastic gradient descent using loss function in (15.25)
14:         Update the actor policy parameter in (15.27)
15:         Update the target networks as in (15.28) and (15.29)
16:         Update the state $s^t = s^{t+1}$
17:     **end for**
18: **end for**

## 15.5 Parallel DRL for joint power allocation and phase-shift matrix Optimisation

In practical applications, when we process all the data in a centralised manner, the information of the UAV's power and the RIS's phase shift for the next action need to transfer at the beginning of each time step. The delay will occur and make the system unable to deal efficiently with the dynamic environment. Thus, we propose a parallel DRL (PDRL) technique for joint power allocation and phase-shift matrix optimisation. As per the definition of the DRL model, the agents do not know the environmental factor. Thus, in our system, the $n$th UAV has no idea about the power of the $m$th UAV and the diagonal matrix at the RIS. Similarly, the RIS controller does not know about the transmit power at the UAV.

To make the UAV and the RIS work cooperatively, we consider multi-agent learning for our system. In particular, each UAV acts as an agent, and the RIS is a separate agent. For all the agents, we define the state space as $\mathscr{S} = \{H_{1,RIS} \Phi\, h_{RIS,11}, H_{1,RIS} \Phi\, h_{RIS,12}, \ldots, H_{n,RIS} \Phi\, h_{RIS,nm}, \ldots, H_{N,RIS} \Phi\, h_{RIS,NM}\}$ with respect to the channel state information, i.e. the compound of channel gains and phase shifts of RIS. The UAV and the RIS process independently; thus, the action space for the $n$th UAV agent is the transmit power $\mathscr{A}_n = \{P_n\}$ and for the RIS agent is the phase-shift matrix $\mathscr{A}_{RIS} = \{\theta_1, \theta_2, \ldots, \theta_K\}$. With the rewards function, we rely on (15.24).

In Algorithm 15.2, with the parallel learning, we have $N+1$ policy for the $N$ UAVs and 1 policy for the RIS in the P-PPO algorithm. In time step $t$, the $n$th UAV decides the transmit power $P_n$, and the RIS chooses the proper phase-shift matrix $\Phi^t$ at the state $s^t$ for maximising the EE performance. In particular, our parallel model is described as in Figure 15.2. The UAV and the RIS have the local information and interact with the environment to search for an optimal policy $\pi^*$. The agents choose and execute the action toward the environment at each time step. Then, the environment will respond with a value of reward towards the agents. Based on the response, the agents adjust the value of parameters in the action-chosen scheme for finding an optimal policy $\pi^*$. The details of our proposed techniques for joint optimisation of power allocation at the UAV and phase-shift matrix at the RIS are described in Algorithm 15.2. The agent $N+1$ represents the RIS controller.

## 15.6 Proximal policy optimisation for centralised and decentralised problem

Instead of using a hybrid model for continuous action space as in the DDPG algorithm, we propose an on-policy algorithm, namely proximal policy optimisation (PPO), with an efficient learning technique to achieve better performance. In the PPO algorithm, we compare the current policy and the obtained policy to find the maximisation of the objective function as:

$$\mathscr{L}(s,a;\theta) = \mathbb{E}\left[\frac{\pi(s,a;\theta)}{\pi(s,a;\theta_{old})}A^\pi(s,a)\right] = \mathbb{E}\left[p_\theta^t A^\pi(s,a)\right], \qquad (15.31)$$

**Algorithm 15.2** Parallel learning for joint power allocation and phase-shift matrix in RIS-assisted UAV communications

1: **for** Agent $\varpi = 1, \ldots, N, N+1$ **do**
2:     Initialise the critic network $Q_\varpi(s, a; \theta_q)$, the target critic networks $Q'_\varpi$ and actor-network $\mu_\varpi(s; \theta_\mu)$, target actor network $\mu'_\varpi$ for the agent $\varpi$
3:     Initialise replay memory pool $\mathcal{D}_\varpi$ for the agent $\varpi$
4: **end for**
5: **for** episode $= 1, \ldots, E$ **do**
6:     Initialise an action exploration process $\mathcal{N}$
7:     Receive initial observation state $s^0$
8:     **for** iteration $= 1, \ldots, T$ **do**
9:         **for** Agent $\varpi = 1, \ldots, N, N+1$ **do**
10:             Execute the action $a^t_\varpi$ obtained at state $s^t$
11:             Update the reward $r^t_\varpi$ according to (15.24)
12:             Observe the new state $s^{t+1}_\varpi$
13:             Store transition $(s^t_\varpi, a^t_\varpi, r^t_\varpi, s^{t+1}_\varpi)$ into replay buffer $\mathcal{B}_\varpi$
14:             Sample randomly a mini-batch of $D$ transitions $(s^i_\varpi, a^i_\varpi, r^i_\varpi, s^{i+1}_\varpi)$ from $\mathcal{B}_\varpi$
15:             Update critic parameter by SGD using the loss (15.25)
16:             Update the actor policy parameter (15.27)
17:             Update the target networks as in (15.28) and (15.29)
18:             Update the state $s^t_\varpi = s^{t+1}_\varpi$
19:         **end for**
20:     **end for**
21: **end for**

where $p^t_\theta = \frac{\pi(s, a; \theta)}{\pi(s, a; \theta_{old})}$ denote the probability ratio and $A^\pi(s, a) = Q^\pi(s, a) - V^\pi(s)$ is an estimator of the advantage function defined in [34]. We use SGD for training networks with a mini-batch $D$ to maximise the objective. Thus, the policy is updated by:

$$\theta^{t+1} = \arg\max \mathbb{E}\left[\mathcal{L}(s, a; \theta^t)\right]. \tag{15.32}$$

In this work, we use the clipping method function clip $(p^t_\theta, 1 - \varepsilon, 1 + \varepsilon)$ for limiting the objective value to avoid the excessive modification as follows [34]:

$$\mathcal{L}^{CLIP}(s, a; \theta) = \mathbb{E}\left[\min\left(p^t_\theta A^\pi(s, a),\ \text{clip}(p^t_\theta, 1 - \varepsilon, 1 + \varepsilon) A^\pi(s, a)\right)\right], \tag{15.33}$$

where $\varepsilon$ is a small constant. We use the upper bound with $1 + \varepsilon$ when the advantage $A^\pi(s, a)$ is positive. In this case, the objective is equal to:

$$\mathcal{L}^{CLIP}(s, a; \theta) = \min\left(\frac{\pi(s, a; \theta)}{\pi(s, a; \theta_{old})}, (1 + \varepsilon)\right) A^\pi(s, a). \tag{15.34}$$

While the advantage $A^\pi(s, a)$ is positive, the minimum term puts a ceiling to the increased objective. Once $\pi(s, a; \theta) > (1 + \varepsilon)\pi(s, a; \theta_{old})$, the objective is limited

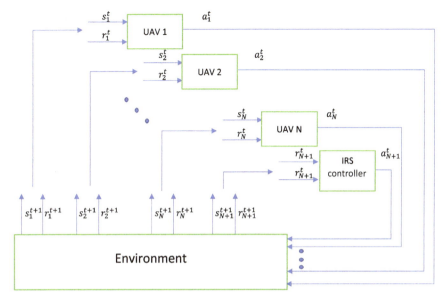

Figure 15.2  A multi-agent learning for the RIS-assisted wireless networks. ©IEEE 2022. Reprinted with permission from [1]

by $(1+\varepsilon)A^\pi(s,a)$. Similarly, when the advantage is negative, the objective can be written as follows:

$$\mathscr{L}^{\text{CLIP}}(s,a;\theta) = \max\left(\frac{\pi(s,a;\theta)}{\pi(s,a;\theta_{old})}, (1-\varepsilon)\right)A^\pi(s,a). \quad (15.35)$$

When the advantage is negative if $\pi(s,a;\theta)$ decreases, the objective will increase. Thus, the maximum term puts a ceiling and once $\pi(s,a;\theta) < (1-\varepsilon)\pi(s,a;\theta_{old})$, the objective is limited by $(1-\varepsilon)A^\pi(s,a)$. These clipping surrogate methods restrict the new policy from going far from the old policy.

Furthermore, we use an advantage function $A^\pi(s,a)$ as follows [35]:

$$A^\pi(s,a) = r^t + \zeta V^\pi(s^{t+1}) - V^\pi(s^t). \quad (15.36)$$

The PPO algorithm is an on-policy algorithm where the UAVs' power level and the RIS's phase-shift matrix value are chosen by running the current policy $\pi(s,a;\theta)$ to maximise the EE performance in (15.24). In our chapter, we use the clipping method in (15.33). For each iteration in the PPO algorithm, a set of trajectory $\mathscr{D} = \{\tau_i\}$ are collected by running the current policy $\pi(s,a;\theta)$ in the environment. Then, the policy parameters are updated by running SGD with Adam optimiser.

## 15.7  Simulation results

We deploy the centralised and parallel learning based on the PPO algorithm, namely, centralised PPO (C-PPO) and parallel PPO algorithm (P-PPO). In the C-PPO algorithm, a policy $\pi(s, a; \theta)$ is used and trained to maximise the EE performance while we have $N + 1$ policy for the $N$ UAVs and for the RIS in the P-PPO algorithm.

For implementing our algorithms, we use Tensorflow 1.13.1 [36]. In our chapter, we consider a scenario with 3 UAVs, $N = 3$ to serve 3 clusters at the fixed location $(0, 0, 200)$, $(200, 300, 200)$, $(400, 0, 200)$. In each cluster, the number of UEs is up to 10. Moreover, we assume $d/\alpha = 1/2$ for convenience. The total power consumption at the RIS and non-transmit power of UAV is set to $P_K + P_c = 4W$. For the neural network, we run the algorithm with different values of parameters and choose the the best performance with the learning rate $lr1 = 0.001$ and $lr2 = 0.002$ for the actor and critic network in the DDPG algorithm, respectively. In the PPO algorithm, we use learning rate $lr = 0.00001$. Other parameters are provided in Table 15.1. In this section, the four proposed schemes in previous sections are summarised as follows:

- **Our centralised DDPG algorithm (C-DDPG)**: As we explained in Section 15.4, we use the DDPG algorithm for jointly optimising the transmit power of the UAV and the phase-shift matrix of the RIS in a centralised manner.
- **Parallel learning for the DDPG method (P-DDPG)**: We consider parallel learning to help to reduce the information transmission delay and errors while ensuring the network performance.
- **Our centralised PPO algorithm (C-PPO)**: Instead of using the DDPG algorithm, we use the PPO algorithm for solving the centralised problem.
- **Parallel learning for the PPO algorithm (P-PPO)**: We deploy the PPO algorithm for parallel learning for joint power allocation and phase-shift matrix optimisation in multi-UAV and RIS-assisted wireless networks.

In addition, to highlight the advantages of our proposals, we also compare our four proposed methods with the following schemes:

- **Max power transmission (MPT)**: We use the maximal transmit power at the UAV and optimise the RIS phase shift by using the PPO algorithm.

*Table 15.1  Simulation parameters*

| Parameters | Value |
| --- | --- |
| Bandwidth ($W$) | 1 MHz |
| UAV transmission power | 5 W |
| UAV's coverage | 500 m |
| The RIS's position | (500, 500, 30) |
| Path-loss parameter | $\kappa_1 = 2, \kappa_2 = 2.2$ |
| Channel power gain | $\beta_0 = -30$ dB |
| Rician factor | $\beta_1 = 4$ |
| Noise power | $\alpha^2 = -134$ dBm |
| Discounting factor | $\zeta = 0.9$ |
| Max number of UEs | 30 |
| Initial batch size | $D = 32$ |

- **Random selection scheme (RSS)**: We randomly select the phase shift at the RIS and optimise the transmit power at the UAV.

Our proposed approaches achieve a better performance in comparison with MPT and RSS methods. Moreover, using the neural networks, the processing time is small, and the UAV and RIS can choose the power allocation and phase-shift matrix immediately in milliseconds. Particularly, in Figure 15.3, we show the EE performance of our proposed method in both centralised and decentralised learning with $M = 10$ and $K = 20$. The methods based on parallel learning reach the best results with the P-PPO algorithm. It is higher than the ones using the C-DDPG and C-PPO algorithms in centralised learning. The convergence of the P-PPO is the fastest and is followed by the P-DDPG algorithm. As can be observed from this figure, our proposed scheme with joint optimisation using the DRL techniques outperforms the other approaches using the MPT and RSS methods.

In Figure 15.4, the EE performance of our methods in comparison with other baseline schemes is presented with the different number of UEs in each cluster, $M$, for the number of RIS elements $K = 20$. Again, the P-PPO method shows better EE performance than the centralised C-PPO and the ones using the C-DDPG algorithm. The reason is that the P-PPO algorithm uses parallel learning and efficient sampling techniques. Moreover, in the DDPG algorithm, we need to use a variable to make randomness to the action for *exploration* and *exploitation*. The MPT and RSS methods are less effective for the joint power allocation and phase shift matrix optimisation in the UAV-assisted wireless network with the support of the RIS.

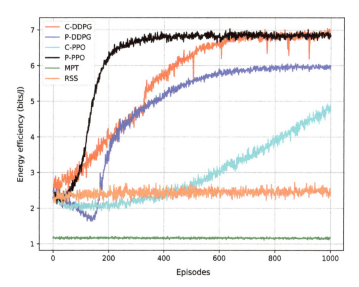

*Figure 15.3*   *EE with $M = 10$ and $K = 20$. ©IEEE 2022. Reprinted with permission from [1]*

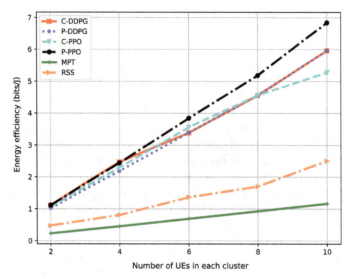

*Figure 15.4  EE versus the number of UEs in each cluster, M. ©IEEE 2022. Reprinted with permission from [1]*

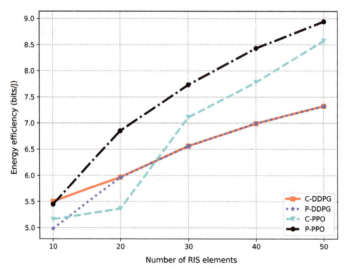

*Figure 15.5  The EE versus the number of the RIS elements, K. ©IEEE 2022. Reprinted with permission from [1]*

In Figure 15.5, we plot the EE performance versus the number of the RIS elements ($K$) when the number of UEs in each cluster equals ten ($M = 10$). We achieve the best EE performance with the P-PPO algorithm despite the value of $K$. When the number of RIS elements becomes higher (e.g. $K > 25$), the methods based on the C-PPO algorithm are more effective than those using the DDPG algorithm. In

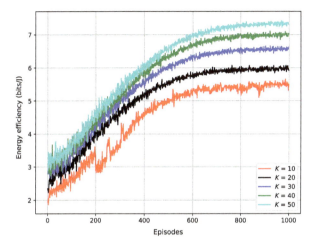

*Figure 15.6  The EE of the C-DDPG algorithm with different number of the RIS elements, K. ©IEEE 2022. Reprinted with permission from [1]*

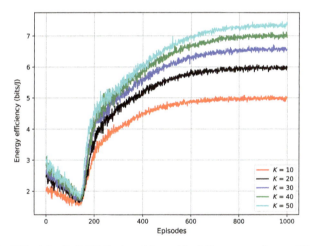

*Figure 15.7  EE of the P-DDPG algorithm with different numbers of RIS elements, K. ©IEEE 2022. Reprinted with permission from [1]*

contrast, for a smaller value of $K$, the methods based on the C-DDPG algorithm are better than the centralised learning with the C-PPO algorithm. For all values of $K$, the best performance can be achieved with the P-PPO algorithm, which demonstrates the fact that the P-PPO algorithm is stable and practical for every environmental setting under the joint optimisation of power allocation at UAVs and phase-shift matrix at RIS.

The EE performances of the DDPG algorithm versus episodes for different numbers of RIS elements using the centralised learning and parallel learning are shown in Figure 15.6 and Figure 15.7, respectively. With the higher number of RIS elements, the performance increases while the convergence rate is still similar for both

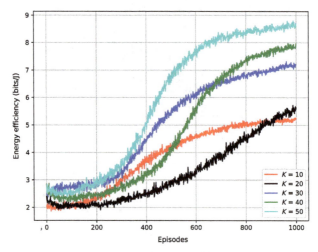

Figure 15.8 *The EE of the C-PPO algorithm with different numbers of the RIS elements, K. ©IEEE 2022. Reprinted with permission from [1]*

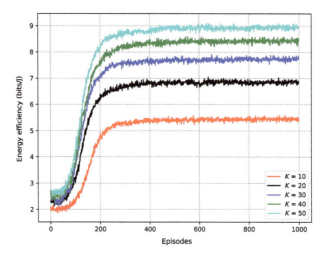

Figure 15.9 *EE of the P-PPO algorithm with different numbers of RIS elements, K. ©IEEE 2022. Reprinted with permission from [1]*

centralised and parallel approaches. The result converges after about 600 episodes when the *exploration* is set to 3 and $\psi = 0.99995$. Thus, depending on the specific purpose, we can deploy the configurable RIS with fast learning.

Similarly, the EE performance of the PPO algorithm versus episodes for different numbers of RIS elements using the centralised learning and parallel earning are plotted in Figure 15.8 and Figure 15.9, respectively. While the performance using centralised approach (C-PPO) is unstable and takes around 800 episodes for convergence, the parallel approach (P-PPO algorithm) shows a solid performance even when increasing the number of the RIS elements. P-PPO convergence is still stable and even faster with the higher number of RIS elements. We need only about 200 episodes for convergence. Furthermore, we use neural networks for the DDPG and PPO algorithm; thus, the system can be easily deployed after training, and the agent can choose the action immediately.

## 15.8 Summary

In this chapter, we have illustrated DRL-based methods, i.e. the DDPG and the PPO techniques, to maximise the EE of multi-UAV networks supported by an RIS. More specifically, the DRL method is used to jointly optimise the transmit power at the UAV and the phase-shift matrix at the RIS. Moreover, to reduce the network's delay and the power to exchange information, we proposed parallel learning for the optimisation problem. The results suggested that we can deploy the DRL algorithms for real-time optimisation with impressive results compared to other baseline schemes.

## References

[1] Nguyen KK, Khosravirad S, Costa DBD, *et al.* Reconfigurable Intelligent Surface-assisted Multi-UAV Networks: Efficient Resource Allocation with Deep Reinforcement Learning. *IEEE J Selected Topics in Signal Process*. 2022;16(3):358–368.

[2] Shakoor S, Kaleem Z, Baig M. I, Chughtai O, Duong T. Q, and Nguyen L. D, "Role of UAVs in public safety communications: Energy efficiency perspective," *IEEE Access*, vol. 7, pp. 140 665–140 679, Sept. 2019.

[3] Vacca A, Onishi H, and Cuccu F. Drones: Military Weapons, Surveillance or Mapping Tools for Environmental Monitoring? The Need for Legal Framework is Required. *Transp Res Proc*. 2017;25:51–62.

[4] Duong TQ, Nguyen LD, Tuan HD, *et al.* Learning-Aided Realtime Performance Optimisation of Cognitive UAV-Assisted Disaster Communication. In: *Proc. IEEE Global Communications Conference (GLOBECOM)*. Waikoloa, HI, USA; 2019.

[5] Nguyen LD, Nguyen KK, Kortun A, *et al.* Real-Time Deployment and Resource Allocation for Distributed UAV Systems in Disaster Relief. In: *Proc. IEEE 20th International Workshop on Signal Processing Advances in Wireless Commun. (SPAWC)*. Cannes, France; 2019. p. 1–5.

[6] Nguyen LD, Kortun A, and Duong TQ. An Introduction of Real-time Embedded Optimisation Programming for UAV Systems under Disaster Communication. *EAI Endorsed Trans Ind Netw Intell Syst*. 2018;5(17):1–8.

[7] Zhan C, Zeng Y, and Zhang R. Energy-Efficient Data Collection in UAV Enabled Wireless Sensor Network. *IEEE Wireless Commun Lett*. 2018;7(3):328–331.

[8] Nguyen KK, Vien NA, Nguyen LD, *et al.* Real-Time Energy Harvesting Aided Scheduling in UAV-Assisted D2D Networks Relying on Deep Reinforcement Learning. *IEEE Access*. 2021;9:3638–3648.

[9] Guo H, Liang YC, Chen J, *et al.* Weighted Sum-Rate Maximization for Reconfigurable Intelligent Surface Aided Wireless Networks. *IEEE Trans Wireless Commun.* 2020;19(5):3064–3076.

[10] Zou Y, Gong S, Xu J, *et al.* Wireless Powered Intelligent Reflecting Surfaces for Enhancing Wireless Communications. *IEEE Trans Veh Technol.* 2020;69(10):12369–12373.

[11] Che Y, Lai Y, Luo S, *et al.* UAV-Aided Information and Energy Transmissions for Cognitive and Sustainable 5G Networks. *IEEE Trans Wireless Commun.* 2021;20(3):1668–1683.

[12] Huang C, Zappone A, Alexandropoulos GC, *et al.* Reconfigurable Intelligent Surfaces for Energy Efficiency in Wireless Communication. *IEEE Trans Wireless Commun.* 2019;18(8):4157–4170.

[13] Basar E, Renzo MD, Rosny JD, *et al.* Wireless Communications Through Reconfigurable Intelligent Surfaces. *IEEE Access.* 2019;7:116753–116773.

[14] Yuan J, Liang YC, Joung J, *et al.* Intelligent Reflecting Surface-Assisted Cognitive Radio System. *IEEE Trans Commun.* 2021;69(1):675–687.

[15] Atapattu S, Fan R, Dharmawansa P, *et al.* Reconfigurable Intelligent Surface Assisted Two–Way Communications: Performance Analysis and Optimization. *IEEE Trans Commun.* 2020;68(10):6552–6567.

[16] Yu H, Tuan HD, Nasir AA, *et al.* Joint Design of Reconfigurable Intelligent Surfaces and Transmit Beamforming Under Proper and Improper Gaussian Signaling. *IEEE J Select Areas Commun.* 2020;38(11):2589–2603.

[17] Ge L, Dong P, Zhang H, *et al.* Joint Beamforming and Trajectory Optimization for Intelligent Reflecting Surfaces-Assisted UAV Communications. *IEEE Access.* 2020;8:78702–78712.

[18] Li S, Duo B, Yuan X, *et al.* Reconfigurable Intelligent Surface Assisted UAV Communication: Joint Trajectory Design and Passive Beamforming. *IEEE Wireless Commun Lett.* 2020;9(5):716–720.

[19] Ranjha A and Kaddoum G. URLLC Facilitated by Mobile UAV Relay and RIS: A Joint Design of Passive Beamforming, Blocklength, and UAV Positioning. *IEEE Internet Things J.* 2021;8(6):4618–4627.

[20] Nguyen KK, Duong TQ, Vien NA, *et al.* Non-Cooperative Energy Efficient Power Allocation Game in D2D Communication: A Multi-Agent Deep Reinforcement Learning Approach. *IEEE Access.* 2019;7:100480–100490.

[21] Nguyen KK, Duong TQ, Vien NA, *et al.* Distributed Deep Deterministic Policy Gradient for Power Allocation Control in D2D-Based V2V Communications. *IEEE Access.* 2019;7:164533–164543.

[22] Feng K, Wang Q, Li X, *et al.* Deep Reinforcement Learning Based Intelligent Reflecting Surface Optimization for MISO Communication Systems. *IEEE Wireless Commun Lett.* 2020;9(5):745–749.

[23] Zeng Y and Zhang R. Energy-Efficient UAV Communication with Trajectory Optimization. *IEEE Trans Wireless Commun.* 2017;16(6):3747–3760.

[24] Yan W, Yuan X, He ZQ, *et al.* Passive Beamforming and Information Transfer Design for Reconfigurable Intelligent Surfaces Aided Multiuser MIMO Systems. *IEEE J Select Areas Commun.* 2020;38(8):1793–1808.

[25] Di B, Zhang H, Song L, *et al*. Hybrid Beamforming for Reconfigurable Intelligent Surface Based Multi-User Communications: Achievable Rates with Limited Discrete Phase Shifts. *IEEE J Select Areas Commun*. 2020;38(8): 1809–1822.

[26] Yu Y, Wang T, and Liew SC. Deep-Reinforcement Learning Multiple Access for Heterogeneous Wireless Networks. *IEEE J Select Areas Commun*. 2019;37(6):1277–1290.

[27] Sheen B, Yang J, Feng X, *et al*. A Deep Learning Based Modeling of Reconfigurable Intelligent Surface Assisted Wireless Communications for Phase Shift Configuration. *IEEE Open J Commun Soc*. 2021;2:262–272.

[28] Huang C, Mo R, and Yuen C. Reconfigurable Intelligent Surface Assisted Multiuser MISO Systems Exploiting Deep Reinforcement Learning. *IEEE J Select Areas Commun*. 2020;38(8):1839–1850.

[29] Shokry M, Elhattab M, Assi C, *et al*. Optimizing Age of Information Through Aerial Reconfigurable Intelligent Surfaces: A Deep Reinforcement Learning Approach. *IEEE Trans Veh Technol*. 2021;70(4):3978–3983.

[30] Bertsekas DP. *Dynamic Programming and Optimal Control*. vol. 1. Athena Scientific Belmont, MA; 1995.

[31] Sutton RS, McAllester D, Singh S, *et al*. Policy Gradient Methods for Reinforcement Learning with Function Approximation. In: *Adv. Neural Inf. Process. Syst.*; 2000. p. 1057–1063.

[32] Lillicrap T. P, Hunt J. J, Pritzel A, *et al*. "Continuous control with deep reinforcement learning," in *Proc. 4th International Conf. on Learning Representations (ICLR)*, 2016.

[33] Kingma DP and Ba JL. *Adam: A Method for Stochastic Optimization*; 2014. Available from: https://arxiv.org/abs/1412.6980.

[34] Schulman J, Moritz P, Levine S, *et al*. High-Dimensional Continuous Control Using Generalized Advantage Estimation. In: *Proc. 4th International Conf. Learning Representations (ICLR)*; 2016.

[35] Mnih V, Badia A. P, Mirza M, *et al*. "Asynchronous methods for deep reinforcement learning," in *Proc. Int. Conf. Mach. Learn. PMLR*, 2016, pp. 1928–1937.

[36] Abadi M, Barham P, Chen J, *et al*. "Tensorflow: A system for large-scale machine learning," in *Proc. 12th USENIX Sym. Opr. Syst. Design and Imp. (OSDI 16)*, Nov. 2016, pp. 265–283.

# Index

action value estimation approaches 41–2
actor-critic learning 73
   different categories of methods 76
      asynchronous advantage actor-critic (A3C) 77
      deep deterministic policy gradient (DDPG) 76–7
      proximal policy optimisation 77–8
   policy gradient 73–4
   reducing variance with the advantage function 75
AdaGrad 26
Adaline activation function 22–3
Adam optimiser 250
adaptive gradient descent 26
adaptive linear neuron model 22–3
additive white Gaussian noise (AWGN) 123
advantage actor-critic (A2C) algorithm 75
agent vs. environment 38–9
air-to-air (A2A) links 198
air-to-air (AA) channels 245
air-to-ground (A2G) links 198
AlphaGo 120
angle of departure (AoD) 245
artificial intelligence (AI) 3–4
artificial neuron 15–16
asynchronous advantage actor-critic (A3C) 75, 77

batch gradient descent 24
Bellman's equation 37–8, 43–4, 55, 103, 146
bias 31
block coordinate descent (BCD) procedure 242

centralised DDPG algorithm (C-DDPG) 254
centralised PPO (C-PPO) algorithm 253–4
channel state information (CSI) 142, 200, 204, 242
classification concept 6
clipping surrogate method 150
clustering 7–8
continuous state-action space, value function approximation for 53
   comparison of different networks 60–1
   deep Q-learning (DQL) 54–6
   double DQL 58
   dueling DQL 58–60
   replay buffer, methods for sampling 56–8
   from tabular to function approximation method 53–4
convolutional neural networks (CNN) 17
COVID-19 pandemic 161

deep deterministic policy gradient (DDPG) algorithm 76–7, 100, 102, 106, 112–13, 142, 152–6, 164, 221, 249–50
deep double Q-learning (DDQL) 128
deep learning (DL) 3–4, 99
deep neural networks (DNNs) 15, 122
   adjusting the coefficients of, through error backpropagation 26
   multi-layer neural network 27–30

deep neural networks (DNNs)
(*continued*)
    artificial neuron 15–16
    learning algorithm 21
        adaptive linear neurons and gradient descent learning method 22–4
        batch gradient descent 24
        gradient descent and SGD, trade-off between 25
        large training datasets, dealing with 24
        learning rate, adjusting 26
        perceptron learning rule 21–2
        stochastic gradient descent (SGD) with momentum 25–6
    multi-layer neural networks, choosing activation functions for 17
        hyperbolic tangent function 18–19
        rectified linear unit (ReLU) 19–21
        sigmoid function 18
        softmax function 21
    neural network architectures 16–17
    overfitting and underfitting 30–2
deep Q-learning (DQL) 54–6, 127–8, 141, 146, 165
    double DQL 58
    dueling DQL 58–60
deep Q-network (DQN) 127
deep reinforcement learning (DRL) 10, 120, 122, 140, 162, 172–4, 204, 218, 228–9, 242
    for power allocation optimisation in U2U communication 127
        deep Q-learning 127–8
        double deep Q-learning 128
        dueling deep Q-learning 128–31
deterministic policy gradient (DPG) 70–1, 105

device-to-device (D2D)
    communications 97–8, 106–9, 140
    RIS-assisted 203
        joint optimisation of power allocation and phase shift matrix 207–9
        simulation results 209–14
        system model and problem formulation 205–7
digital twin (DT) paradigm 89, 200
    deep reinforcement learning 88
    resource management in UAV-enabled communications 89
    trajectory planning for UAV 90
dimensionality reduction 9–10
distributed deep deterministic policy gradient (DDDPG) 97, 100, 106, 110–14
    multi-agent power allocation problem
        in D2D-based V2V communications 106–9
        in U2U-based communications 103, 105–6
    simulation results 110–15
    system model and problem formulation 100–3
double deep Q-learning (DDQL) 58, 128, 141
DRL 97, 99
dueling deep Q-learning 128–31
dynamic programming (DP) 43–6

effective deep reinforcement learning approach for UAV-assisted IoT networks 170–2
efficient energy management 87–8
electromagnetic (EM) metamaterials 194
energy efficiency (EE) performance 85, 100, 203

energy harvesting time scheduling in UAV-powered D2D communications 148–9
energy receivers (ERs) 197
epsilon-greedy method 40
error backpropagation, adjusting the deep neural network's coefficients through 26–8
every-visit 47
exploration and exploitation 12, 39–40
exponential linear unit (ELU) 20

feedforward neural networks (FNN) 17
field-programmable gate array (FPGA) 194
first-visit 47
fixed interval update 56

gated recurrent units (GRU) 17
Gaussian exploration policy 106
general policy iteration (GPI) 46
generative adversarial networks (GAN) 17
gradient descent 23
  adaptive 26
  batch 24
  mini-batch 25
  stochastic 24–6, 30, 76, 209, 227, 249
gradient descent and SGD, trade-off between 25

hybrid sampling methods 58
hyperbolic tangent function 18–19

information receivers (IRs) 197
Internet of Things (IoT) 83
iterative policy evaluation 43

key performance indicators (KPIs) 193
K-means clustering algorithm 7–8, 141

Kullback–Leibner (KL) divergence 66, 142, 150–1

Lagrangian relaxation 66
large training datasets, dealing with 24
learning algorithm 21
  adaptive linear neurons and gradient descent learning method 22–4
  batch gradient descent 24
  gradient descent and SGD, trade-off between 25
  large training datasets, dealing with 24
  learning rate, adjusting 26
  perceptron learning rule 21–2
  stochastic gradient descent (SGD) with momentum 25–6
learning rate, adjusting 26
line-of-sight (LoS) links 166, 223
long short-term memory networks (LSTM) 17

machine learning (ML) 3–5, 162
Markov decision process (MDP) 35–6, 124, 145, 166, 204, 221, 247
  dynamic programming (DP) 43–6
  Monte Carlo methods 46
    estimation procedures 47–8
    general policy iteration in 48–9
  multi-arm bandit (MAB) problem 39
    action value estimation approaches 41–2
    epsilon-greedy method 40
    exploration and exploitation 39–40
    softmax selection 40–1
  reinforcement learning (RL) 35
    agent vs. environment 38–9
    Bellman's equation 37–8
    policy classification and expected reward 36–7

Markov decision process (MDP) (*continued*)
   temporal difference (TD) learning 49
      estimation procedures 50
      general policy iteration 50–1
maximal power transmission (MPT) 209, 232, 254
McCulloch–Pitt (MCP) neuron model 15, 22
mini-batch gradient descent algorithm 25
mobile edge computing (MEC) paradigm 197
model-based RL 11, 38
model-free learning algorithms 11
model-free RL 38
Monte-Carlo methods 46, 48, 248
   estimation procedures 47–8
   general policy iteration in 48–9
Monte Carlo simulation 64, 73, 105
multi-agent learning in networks supported by RIS and multi-UAVs 241
   contributions 243–4
   parallel DRL for joint power allocation and phase-shift matrix optimisation 251
   power allocation and phase-shift matrix, centralised optimisation for 248–50
   preliminaries 247
      policy search 248
      value function 247
   proximal policy optimisation for centralised and decentralised problem 251–3
   related work 242–3
   simulation results 253–9
   system model and problem formulation 244
      problem formulation 246–7
      system model 244–6

multi-agent power allocation problem in U2U-based communications 103
   distributed deep deterministic policy gradient 105–6
   policy search 104–5
   value function 103–4
multi-agent Q-learning approach 125–7
multi-arm bandit (MAB) problem 39
   action value estimation approaches 41–2
   epsilon-greedy method 40
   exploration and exploitation 39–40
   softmax selection 40–1
multicell networks, RIS-assisted 196
multi-layer neural networks 27–30
   choosing activation functions for 17
      hyperbolic tangent function 18–19
      rectified linear unit (ReLU) 19–21
      sigmoid function 18
      softmax function 21
multiple-input multiple-output (MIMO) 122, 194
multiple-input single-output (MISO) 242

natural policy gradient methods 65
   proximal policy optimisation 68–70
   trust region policy optimisation 67–8
Nelder–Mead simplex 219
Nesterov accelerated gradient (NAG) algorithm 25–6
neural network architectures 16–17
next-generation wireless networks, reconfigurable intelligent surface (RIS) in 195
   mobile edge computing networks, RIS-assisted 197

multicell networks, RIS-assisted 196
non-orthogonal multiple access, RIS-aided 196
physical layer security, RIS for 197
simultaneous wireless information and power transfer, RIS for 196–7
non-cooperative energy-efficient power allocation game in U2U communication 119
   deep reinforcement learning for power allocation optimisation 127
      deep Q-learning 127–8
      double deep Q-learning 128
      dueling deep Q-learning 128–31
   reinforcement learning for energy-efficient power allocation game 124
      multi-agent Q-learning approach 125–7
      single-agent Q-learning 124–5
   related work 121–3
   simulation results 131
      exploration/exploitation analysis 132–4
      performance comparison 132
      running time analysis 134–5
      scalability analysis 132
   system model and problem formulation 123–4
non-orthogonal multiple access (NOMA) technique 122, 196

offline learning 38
off-policy methods 49
online learning 24, 38
on-policy methods 49
OpenAI 68, 75
optimal harvesting time optimisation (OHT) solution 151
overfitting 30–2
overshooting 65

parallel deep reinforcement learning (PDRL) technique 251
parallel DRL for joint power allocation and phase-shift matrix optimisation 251
parallel learning for the DDPG method (P-DDPG) 254
parallel learning for the PPO algorithm (P-PPO) 253–4
perceptron learning rule 21–2
physical layer security (PLS) techniques 197
policy evaluation 43
policy gradient 64, 73–4
policy iteration 44
policy optimisation algorithms 221
   efficient learning with 150
      clipping surrogate method 150
      Kullback–Leibler divergence penalty 150–1
policy search 147–8
policy search methods for reinforcement learning 63
   deterministic policy gradient 70–1
   natural policy gradient methods 65
   proximal policy optimisation 68–70
   trust region policy optimisation 67–8
   policy gradient 64
   rationale behind policy search 63
   REINFORCE algorithm 64–5
power allocation and phase-shift matrix, centralised optimisation for 248–50
principal component analysis (PCA) 9
prioritised experience replay sampling 57
proximal policy optimisation (PPO) 67–70, 77–8, 142, 164, 208, 213, 229–31, 251–3

Q-learning 51
  deep 127–8
  double deep 128
  dueling deep 128–31
  multi-agent 125–7
  single-agent 124–5
Q-table 12–13
quality-of-service (QoS) constraint 144
$Q$-value function 53–5, 59, 70–1

random phase shift matrix selection (RPS) 210
random selection scheme (RSS) 232, 255
rationale behind policy search 63
real-time energy harvesting-aided scheduling in UAV-assisted D2D networks 139
  contributions 142
  deep deterministic policy gradient approach 148–9
  preliminaries 146
    policy search 147–8
    value function 146–7
  proximal policy optimisation algorithms, efficient learning with 150
    clipping surrogate method 150
    Kullback–Leibler divergence penalty 150–1
  simulation results 151
    computational complexity 156–7
    parameter analysis 153–6
    performance comparison 151–3
  state of the art and challenges 140–2
  system model and problem formulation 142–5
reconfigurable intelligent surface (RIS) 194–5, 204, 218, 241
  -assisted D2D communications 203
    joint optimisation of power allocation and phase shift matrix 207–9
    simulation results 209–14
    system model and problem formulation 205–7
  -assisted UAV communications 217
    channel model 222–3
    contributions 220–1
    hovering UAV for downlink power transfer and uplink information transmission in 225
    information transmission phase 224–5
    joint trajectory, transmission power, energy harvesting time scheduling, and RIS phase shift optimisation 228–9
    organisation 221
    power transfer phase 223–4
    proximal policy optimisation technique 229–31
    simulation results 231–7
    state of the art 218–20
    system model and problem formulation 221–4
  challenges and research directions ahead 199
    channel state estimation techniques 199–200
    mobility management 200
    RIS deployment strategies 200
    use of AI in RIS-assisted communications 200–1
  empowered UAV-assisted communications 198
    air-to-ground (A2G) links 198
    RIS-equipped UAV communications 198–9
  in next-generation wireless networks 195
    mobile edge computing networks, RIS-assisted 197

multicell networks, RIS-assisted 196
non-orthogonal multiple access, RIS-aided 196
physical layer security, RIS for 197
simultaneous wireless information and power transfer, RIS for 196–7
*see also* multi-agent learning in networks supported by RIS and multi-UAVs
rectified linear unit (ReLU) 19–21
recurrent neural networks (RNN) 17
regression models 6–7
REINFORCE algorithm 64–5, 73, 105, 248
reinforcement learning (RL) algorithms 10, 35, 120
  agent vs. environment 38–9
  Bellman's equation 37–8
  for energy-efficient power allocation game in U2U communication 124
    multi-agent Q-learning approach 125–7
    single-agent Q-learning 124–5
  environment 10–11
  policy 12–13
  policy classification and expected reward 36–7
  reward 11–12
  training of 13
reinforcement learning, policy search methods for 63
  deterministic policy gradient 70–1
  natural policy gradient methods 65
    proximal policy optimisation 68–70
    trust region policy optimisation 67–8
  policy gradient 64
  rationale behind policy search 63
  REINFORCE algorithm 64–5
relative entropy 66

resource management in UAV-enabled communications 89
Rician factor 245
Rician fading channel 222
Rosenblatt's perceptron model 22–3, 26

sample-average method 41
sharing deep deterministic policy gradient (SDDPG) 100, 106, 108, 114
sigmoid function 18
signal-to-interference-plus-noise ratio (SINR) 101, 123, 144, 218, 246
signal-to-noise ratio (SNR) 84, 195, 220, 243
simultaneous wireless information and power transfer (SWIPT) 197, 203
single-agent Q-learning 124–5
6G networks requirements 83
  guaranteed throughput 85
  system efficiency 85
softmax function 21, 40–1
sparse rewards 11
spatial division multiple access (SDMA) 196
stochastic gradient descent (SGD) 24–6, 30, 76, 209, 227, 249
sum of squared error (SSE) 23
supervised learning 5
  classification concept 6
  regression models 6–7
system efficiency (SE) 85

tanh function: *see* hyperbolic tangent function
target network update interval 56
target update rate 56
Taylor expansion 66
temporal difference (TD) learning 49
  estimation procedures 50
  general policy iteration in 50–1

Third Generation Partnership Project
(3GPP) 85
3D path planning 88
3D trajectory design and data
collection in UAV-assisted
networks 161
  contributions and organisation
    163–5
  deep reinforcement learning
    approach for UAV-assisted
    IoT networks 172–4
  effective deep reinforcement
    learning approach for
    UAV-assisted IoT networks
    170–2
  preliminaries 169–70
  related contributions 162–3
  simulation results 174
    expected reward 174–8
    parametric study 181–6
    throughput comparison 178–81
  system model and problem
    formulation 165
    game formulation 166–9
    observation model 165–6
trajectory planning for UAV 90
trust region policy optimisation
(TRPO) 67–8

UAV-to-UAV (U2U) communication
  deep reinforcement learning for
    power allocation optimisation
    in 127
    deep Q-learning 127–8
    double deep Q-learning 128
    dueling deep Q-learning
      128–31
  multi-agent power allocation
    problem in 103
    distributed deep deterministic
      policy gradient 105–6
    policy search 104–5
    value function 103–4
  reinforcement learning for
    energy-efficient power
    allocation game in 124

multi-agent Q-learning
  approach 125–7
single-agent Q-learning 124–5
ultra-reliable and low-latency
  communication (URLLC)
  83–4, 243
underfitting 30–2
undershooting 65
uniform random sampling 56
unmanned aerial vehicle
  (UAV)-assisted 6G
  communications 83
  benefits of UAV integration 85
  main UAV-assisted services
    86–7
  digital twin and deep
    reinforcement learning 88
  resource management in
    UAV-enabled
    communications 89
  trajectory planning for UAV 90
  open challenges 87
    3D path planning 88
    efficient energy management
      87–8
  6G networks requirements 83
    guaranteed throughput 85
    system efficiency 85
unmanned aerial vehicle
  (UAV)-assisted IoT networks
  170–2
  deep reinforcement learning
    approach for 172–4
  effective deep reinforcement
    learning approach for
    170–2
unmanned aerial vehicle (UAV)
  communications 241
  RIS-equipped 198–9
  *see also* multi-agent learning in
    networks supported by RIS
    and multi-UAVs
unmanned aerial vehicle (UAV)
  communications,
  RIS-assisted 217
  contributions 220–1

hovering UAV for downlink power
    transfer and uplink
    information transmission
    in225
  DDPG method  226–7
  game solving  227–8
  preliminaries  225–6
  joint trajectory, transmission
    power, energy harvesting time
    scheduling, and RIS phase
    shift optimisation  228–9
  organisation  221
  proximal policy optimisation
    technique  229–31
  simulation results  231–7
  state of the art  218–20
  system model and problem
    formulation  221
    channel model  222–3
    information transmission phase
      224–5
    power transfer phase  223–4
unmanned aerial vehicle
    (UAV)-powered D2D
    communications, energy
    harvesting time scheduling in
    148–9
unsupervised learning  7
  clustering  7–8
  dimensionality reduction  9–10

value function  146–7
value function approximation for
    continuous state-action space
    53
  comparison of different networks
    60–1
  deep Q-learning (DQL)  54–6
  double DQL  58
  dueling DQL  58–60
  replay buffer, methods for
    sampling  56–8
  from tabular to function
    approximation method  53–4
value iteration technique  46
vanilla policy gradient method (VPG)
    73, 210
variance  31
vehicle-to-vehicle (V2V)
    communications  98, 122
  device-to-device (D2D)-based
    106–9
visit  47
$V$-value function  53–4

wild dataset  9
wireless information transmission
    (WIT)  218, 222
wireless power transfer (WPT)  87,
    218, 222

www.ingramcontent.com/pod-product-compliance
Lightning Source LLC
Jackson TN
JSHW082151080725
87326JS00003B/38